高等学校智能科学与技术/人工智能专业教材

# 人工智能专业英语教程
## （第2版）

朱 丹 朱宏峰 刘 伟 编著

清华大学出版社

北京

# 内 容 简 介

　　本书是人工智能、计算机、自动控制等相关专业的专业英语教材，选材广泛，内容涵盖人工智能的基本概念、发展历史、主要技术、人工智能的现在与未来，以及人工智能给人类带来的影响和人工智能的应用领域等。本书具体内容包括：第 1 章介绍人工智能的基本概念；第 2 章介绍人工智能的发展历史；第 3 章～第 5 章介绍人工智能的主要技术，包括机器学习、深度学习和自然语言处理等；第 6 章～第 12 章介绍人工智能的应用领域，包括人工智能在农业、教育、安防、金融、医疗、交通、家庭等领域的应用情况；第 13 章介绍人工智能的影响。书中每章所选用文章均来自国外网站，本书作者对文章中出现的新词和专业术语进行了注释，每篇文章都配有相应的习题和拓展阅读，有利于读者巩固学习效果。

　　本书可作为高等院校人工智能、计算机、自动控制等专业的英文课程教材，同时也可作为各类计算机从业人员或者有志于投身人工智能领域的相关人员的自学书籍。

**图书在版编目（CIP）数据**

人工智能专业英语教程 / 朱丹，朱宏峰，刘伟编著. —2 版. —北京：清华大学出版社，2023.1(2024.1 重印)

高等学校智能科学与技术 / 人工智能专业教材

ISBN 978-7-302-61438-8

Ⅰ. ①人…　Ⅱ. ①朱… ②朱… ③刘…　Ⅲ. ①人工智能 - 英语 - 高等学校 - 教材　Ⅳ. ①TP18

中国版本图书馆 CIP 数据核字（2022）第 133661 号

责任编辑：赵　凯
封面设计：常雪影
责任校对：焦丽丽
责任印制：宋　林

出版发行：清华大学出版社
　　　网　　　址：https://www.tup.com.cn，https://www.wqxuetang.com
　　　地　　　址：北京清华大学学研大厦 A 座　　　邮　　　编：100084
　　　社　总　机：010-83470000　　　邮　　　购：010-62786544
　　　投稿与读者服务：010-62776969，c-service@tup.tsinghua.edu.cn
　　　质　量　反　馈：010-62772015，zhiliang@tup.tsinghua.edu.cn
　　　课　件　下　载：https://www.tup.com.cn，010-83470236
印　装　者：北京嘉实印刷有限公司
经　　销：全国新华书店
开　　本：185mm×260mm　　　印　张：12.75　　　字　数：312 千字
版　　次：2020 年 1 月第 1 版　2023 年 3 月第 2 版　　　印　次：2024 年 1 月第 2 次印刷
印　　数：1501～2500
定　　价：69.00 元

产品编号：095949-01

# 第2版 前言
## FOREWORD

人工智能产业是智能产业发展的核心，是其他智能科技产品发展的基础，国内外的高科技公司以及风险投资机构纷纷布局人工智能产业链。前瞻产业研究院发布的《人工智能行业市场前瞻与投资战略规划分析报告》中指出，2017 年中国人工智能核心产业规模超过 700 亿元，随着国家相关规划的出台，各地人工智能相关建设将逐步启动，2020 年，中国人工智能核心产值规模超过 1600 亿元。在全球人工智能人才竞争的大背景下，我国对人工智能高等教育愈发重视。全国已有 700 多所高校围绕人工智能领域设置了二级学科或交叉学科。人工智能从业人员必须提高专业英语水平，以便及时获取最新、最先进的专业知识。因此，该专业的高校都有开设人工智能专业英语课程的需求。

本书选材广泛，内容涵盖人工智能的基本概念、发展历史、主要技术、人工智能的现在与未来，以及人工智能给人类带来的影响和人工智能的应用领域等。本书共分 13 章，第 1 章介绍人工智能的基本概念；第 2 章介绍人工智能的发展历史；第 3～5 章介绍人工智能的主要技术，包括机器学习、深度学习和自然语言处理等；第 6～12 章介绍人工智能的应用领域，包括人工智能在农业、教育、安防、金融、医疗、交通、家庭等领域的应用情况；第 13 章介绍人工智能的影响。书中每个单元包括：Text A 及 Text B 文章，这些文章均选自国外知名网站，具有一定的知识性和实用性；New Words and Expressions 给出课文中出现的新词，读者由此可以扩充词汇量；Terms 对文中出现的专业术语进行解释；Comprehension 针对课文练习，有利于读者巩固学习效果；Answers 给出参考答案，读者可对照检查学习效果；参考译文帮助读者理解文章大意；常用人工智能词汇中英文对照表供读者记忆单词和查询之用。

本书第 1～10 章由朱丹编写，第 11 章、12 章及词汇对照表由朱宏峰编写，第 13 章由刘伟编写。全书由朱丹统稿。

本书文章节选自互联网，在此向文章原作者表示感谢。

由于作者水平有限，书中难免出现不足之处，敬请读者不吝指正。

编者
2022 年 10 月

人工智能产业是智能产业发展的核心，是其他智能科技产品发展的基础，国内外的高科技公司以及风险投资机构纷纷布局人工智能产业链。前瞻产业研究院《人工智能行业市场前瞻与投资战略规划分析报告》指出，2017年中国人工智能核心产业规模超过700亿元，随着国家相关规划的出台，各地人工智能相关建设将逐步启动，预计到2020年，中国人工智能核心产值规模将超过1600亿元。在全球人工智能人才竞争的大背景下，我国对人工智能高等教育愈发重视。截至2017年12月，全国共有71所高校围绕人工智能领域设置了86个二级学科或交叉学科。人工智能从业人员必须提高专业英语水平，以便及时获取最新、最先进的专业知识，因此，该专业的高校都有开设人工智能专业英语课程的需求。

本书选材广泛，内容涵盖人工智能的基本概念、发展历史、主要技术、人工智能的现在与未来，以及人工智能给人类带来的影响和人工智能的应用领域等。本书共分13章，第1章介绍人工智能的基本概念。第2章介绍人工智能的发展历史。第3～5章介绍人工智能的主要技术，包括机器学习、深度学习和自然语言处理等。第6～12章介绍人工智能的应用领域，包括人工智能在农业、教育、安防、金融、医疗、交通、家庭等领域的应用情况。第13章介绍人工智能的影响。书中每个单元包括：Text A及Text B两篇文章，这些文章均选自国外知名网站，具有一定的知识性和实用性；New Words and Expressions给出课文中出现的新词，读者由此可以扩充词汇量；Terms对文中出现的专业术语进行解释；Comprehension针对课文练习，有利于读者巩固学习效果；Answers给出参考答案，读者可对照检查学习效果；参考译文帮助读者理解文章大意；常用人工智能词汇中英文对照表供读者记忆单词和查询之用。

本书第1～10章由朱丹编写，第11章及词汇对照表由蔡丹编写，第12章由王敏编写，第13章由程娟编写。全书由朱丹统稿。

本书文章节选自互联网，在此向文章原作者表示感谢，由于作者水平有限，书中难免出现不足之处，敬请读者不吝指正。

编者
2019年4月

CONTENTS 目 录

# Chapter *1*

# What is Artificial Intelligence

## Text A

*Artificial* Intelligence (AI), sometimes called machine intelligence, is intelligence *demonstrated* by machines, in contrast to the natural intelligence displayed by humans and other animals. In computer science AI research is defined as the study of "intelligent *agents*": any device that *perceives* its environment and takes actions that *maximize* its chance of successfully achieving its goals. Colloquially, the term "artificial intelligence" is applied when a machine *mimics* "*cognitive*" functions that humans *associate* with other human minds, such as "learning" and "problem solving".

For years, it was thought that computers would never be more powerful than the human brain, but as development has *accelerated* in modern times, this has proven to be not the case.

AI as a concept refers to computing hardware being able to *essentially* think for itself, and make decisions based on the data it is being fed. AI systems are often hugely complex and powerful, with the ability to process *unfathomable* depths of information in an extremely quick time in order to come to an effective conclusion.

Thanks to detailed *algorithms*, AI systems are now able to perform *mammoth* computing tasks much faster and more efficiently than human minds, helping making big *strides* in research and development areas around the world.

Some of the most notable real-world applications of AI are IBM's Watson[1], which is being used to power research in a huge

---

### New Words and Expressions

**artificial** /ɑːtɪˈfɪʃ(ə)l/ adj.
人造的；仿造的
**demonstrate** /ˈdemənstreɪt/ vt.
证明；展示；论证
**agent** /ˈeɪdʒənt/ n.
代理；主体
**perceive** /pəˈsiːv/ vt.
察觉，感觉；理解；认知
**maximize** /ˈmæksɪmaɪz/ vt.
取…最大值；对…极为重视
**mimic** /ˈmɪmɪk/ vt.
模仿，摹拟
**cognitive** /ˈkɒɡnɪtɪv/ adj.
认知的，认识的
**associate** /əˈsəʊʃɪeɪt/ vt.
使联合；使发生联系
**accelerate** /əkˈseləreɪt/ vi.
加速；促进；增加
**essentially** /ɪˈsenʃ(ə)li/ adv.
本质上；本来
**unfathomable** /ʌnˈfæð(ə)məb(ə)l/ adj.
深不可测的；无底的；高深莫测的
**algorithm** /ˈælɡərɪð(ə)m/ n.
【计】【数】算法，运算法则
**mammoth** /ˈmæməθ/ adj.
巨大的，庞大的
**stride** /straɪd/ n.
进展

range of fields, with Microsoft's Azure[2] Machine Learning and TensorFlow[3] also making headlines around the world.

But AI-powered smart assistants are becoming a common presence on mobile devices too, with the likes of Siri, Cortana and Alexa all being welcomed into many people's lives.

There seems no limit to the applications of AI technologies, and perhaps the most exciting aspect of the ecosystem is that there's no telling where it can go next, and what problems it may *ultimately* be able to solve.

### Why is Artificial Intelligence Important?

AI *automates* repetitive learning and discovery through data. But AI is different from hardware-driven, robotic automation. Instead of automating manual tasks, AI performs frequent, *high-volume*, computerized tasks *reliably* and without fatigue. For this type of automation, human inquiry is still essential to set up the system and ask the right questions.

AI adds intelligence to existing products. In most cases, AI will not be sold as an individual application. Rather, products you already use will be improved with AI capabilities, much like Siri was added as a feature to a new generation of Apple products. Automation, conversational platforms, *bots* and smart machines can be combined with large amounts of data to improve many technologies at home and in the workplace, from security intelligence to investment analysis.

AI adapts through progressive learning algorithms to let the data do the programming. AI finds structure and *regularities* in data so that the algorithm acquires a skill: The algorithm becomes a classifier or a predictor. So, just as the algorithm can teach itself how to play chess, it can teach itself what product to recommend next online. And the models adapt when given new data. Back *propagation*[4] is an AI technique that allows the model to adjust, through training and added data, when the first answer is not quite right.

AI analyzes more and deeper data using *neural* networks that have many hidden layers. Building a fraud detection[5] system with five hidden layers was almost impossible a few years ago. All that has changed with incredible computer power and big data. You need lots of data to train deep learning models because they learn directly from the data. The more data you can feed them, the more accurate they become.

---

***New Words and Expressions***

**ultimately** /ˈʌltɪmətlɪ/ adv.
　最后；根本；基本上

**automate** /ˈɔːtəmeɪt/ vt.
　使自动化，使自动操作

**high-volume**　adj.
　大容量

**reliably** /rɪˈlaɪəblɪ/ adv.
　可靠地；确实地

**bots** /bɒts/ n.
　机器人

**regularity** /ˌregjʊˈlærətɪ/ n.
　规则性；整齐；正规

**propagation** /ˌprɒpəˈgeɪʃən/ n.
　传播；繁殖；增殖

**neural** /ˈnjʊər(ə)l/ adj.
　神经的；神经系统的

AI achieves incredible accuracy through deep neural networks—which was previously impossible. For example, your interactions with Alexa, Google Search and Google Photos are all based on deep learning—and they keep getting more accurate the more we use them. In the medical field, AI techniques from deep learning, image classification and object recognition can now be used to find cancer on *MRIs* with the same accuracy as highly trained *radiologists*.

AI gets the most out of data. When algorithms are self-learning, the data itself can become intellectual property. The answers are in the data; you just have to apply AI to get them out. Since the role of the data is now more important than ever before, it can create a competitive advantage. If you have the best data in a competitive industry, even if everyone is applying similar techniques, the best data will win.

***New Words and Expressions***
**MRI** abbr.
（Magnetic Resonance Imaging）
核磁共振成像
**radiologist**/ˌreɪdɪˈɑlədʒɪst/ n.
放射科医生

## Terms

### 1. IBM Watson

IBM Watson 是认知计算系统的杰出代表，也是一个技术平台。认知计算代表一种全新的计算模式，它包含信息分析、自然语言处理和机器学习领域的大量技术创新，能够助力决策者从大量非结构化数据中提取和分析重要信息，提升洞察能力。

### 2. Azure

Azure 机器学习服务是一项云服务，可以使用它来训练、部署、自动执行以及管理机器学习模型，所有这些都是在云提供的广泛范围内进行的。

### 3. TensorFlow

TensorFlow 是一个基于数据流编程（dataflow programming）的符号数学系统，被广泛应用于各类机器学习（machine learning）算法的编程实现，其前身是谷歌的神经网络算法库 DistBelief。TensorFlow 拥有多层级结构，可部署于各类服务器、PC 终端和网页并支持 GPU和TPU高性能数值计算，被广泛应用于谷歌内部的产品开发和各领域的科学研究。

### 4. Back Propagation

Back Propagation（反向传播算法）是目前用来训练人工神经网络（Artificial Neural Network，ANN）的最常用且最有效的算法。其主要思想是：

（1）将训练集数据输入 ANN 的输入层，经过隐藏层，最后达到输出层并输出结果，这是 ANN 的前向传播过程；

（2）由于 ANN 的输出结果与实际结果有误差，则计算估计值与实际值之间的误差，将被从输出层向隐藏层反向传播，直至传播到输入层；

（3）在反向传播的过程中，根据误差调整各种参数的值；不断迭代上述过程，直至收敛。

### 5. Fraud Detection

Fraud Detection 为反欺诈中所用到的机器学习模型。

反欺诈应用的机器模型算法，多为二分类算法：

（1）梯度提升决策树（Gradient Boosting Decision Tree，GBDT）算法，该算法的性能高，且在各类数据挖掘中应用广泛，表现优秀，应用场景较多。

（2）Logistic 回归又称 Logistic 回归分析，是一种广义的线性回归分析模型，常用于数据挖掘、疾病自动诊断、经济预测等领域，在有标注样本下训练模型对不同的欺诈情况进行二元判别。

（3）非监督的异常检测的方法，主要是从数据中找出异常的点，这些异常往往与欺诈有关联。

## Comprehension

**Blank Filling**

1. Artificial Intelligence (AI), sometimes called_____, is intelligence demonstrated by machines, in contrast to the _____ intelligence displayed by humans and other animals.

2. AI research is defined as the study of "intelligent agents": any device that perceives its_____ and takes actions that _____ its chance of successfully achieving its goals.

3. The term "artificial intelligence" is applied when a machine mimics "_____" functions that humans associate with other _____, such as "learning" and "problem solving".

4. AI as a concept refers to computing _____ being able to essentially think for itself, and make decisions based on the _____ it is being fed.

5. Thanks to detailed _____, AI systems are now able to perform mammoth _____ tasks much faster and more efficiently than human minds.

6. AI-powered smart assistants are becoming a common presence on _____.

7. AI automates repetitive learning and discovery through _____.

8. AI adapts through _____ to let the data do the programming.

9. AI finds _____ and _____ in data so that the algorithm acquires a skill.

**Content Questions**

1. What is artificial intelligence?

2. In computer science AI research is defined as the study of "intelligent agents". What does "intelligent agents" refer to?

3. How AI is applied in the real world?

4. What is "back propagation"?

## Answers

**Blank Filling**

1. machine intelligence; natural

2. environment; maximize

3. cognitive; human minds

4. hardware; data

5. algorithms; computing

6. mobile devices

7. data

8. progressive learning algorithms

9. structure; regularities

**Content Questions**

1. Artificial Intelligence (AI), sometimes called machine intelligence, is intelligence demonstrated by machines, in contrast to the natural intelligence displayed by humans and other animals.

2. It refers to any device that perceives its environment and takes actions that maximize its chance of successfully achieving its goals.

3. Some of the most notable real-world applications of AI are IBM's Watson, which is being used to power research in a huge range of fields, with Microsoft's Azure Machine Learning and TensorFlow also making headlines around the world.

4. Back propagation is an AI technique that allows the model to adjust, through training and added data, when the first answer is not quite right.

# 参考译文 A

人工智能（AI），有时被称为机器智能。与人类和其他动物展示的天生智能不同，人工智能是由机器展示的智能。在计算机科学中，人工智能研究被定义为对"智能主体"的研究：能够感知环境，并采取行动以成功实现目标的智能机器。一般来说，"人工智能"一词是指机器模仿人类的"认知"功能和"智能"行为，如"学习"和"解决问题"等。

多年来，人们一直认为计算机永远不会比人脑强大，但随着现代科技的发展，事实证明并非如此。

人工智能的概念是指计算硬件能够独立思考，并根据输入的数据做出决策。人工智能系统通常非常复杂和强大，能够在极短的时间内处理深不可测的信息，以便得出有效的结论。

得益于周密的算法，人工智能系统现在能够比人脑更快、更有效地执行庞大的计算任务，从而有助于世界各地的研究和开发取得重大进展。

人工智能最引人注目的一些应用有：IBM 的沃森（Watson）被用于许多领域的研究，微软的 Azure 机器学习（Azure Machine Learning）和 TensorFlow 也都引起了大家的关注。

但是，随着 Siri、Cortana 和 Alexa 等智能助手进入人们的生活，人们的移动设备也应用了人工智能。

人工智能技术的应用似乎没有限制，这正是它令人兴奋的地方，人们不知道它下一步将走向何方，也不知道最终可能解决什么问题。

### 为什么人工智能至关重要？

人工智能通过渐进式学习算法进行调整，让数据进行编程。人工智能在数据中发现结构和规律，使算法获得一种技能：算法成为分类器或预测器。因此，就像算法可以自学如何下棋一样，它也可以自学下一步应该推荐的产品类型。当得到新的数据时，模型就会适应。反向传播是一种人工智能技术，当第一个答案不太正确时，它允许模型通过训练和添加数据来调整。

人工智能通过数据来实现自动化的重复学习和发现。但人工智能不同于硬件驱动的机器人自动化。人工智能不再自动完成手工任务，而是可靠地、不知疲倦地执行频繁的、大批量的计算机任务。对于这种类型的自动化，人工查询对于建立系统和提出正确的问题仍然是必不可少的。

人工智能提高了现有产品的智能化。在大多数情况下，人工智能不会作为一个单独的应用程序出售，而是使在用产品得到改进，就像 Siri 被添加到新一代苹果产品中一样。自动化、对话平台、机器人和智能机器可以与大量数据相结合，使安全智能、投资分析等许多家庭和工作场所的技术得以改进。

人工智能使用具有许多隐藏层的神经网络分析更多和更深层次的数据。多年前，建立一个包含五个隐藏层的欺诈检测系统几乎是不可能的。这一切都因计算机的强大功能和大数据而改变。人们需要大量的数据来训练深度学习模型，因为它们直接从数据中学习。你能提供给它们的数据越多，它们就越准确。

人工智能通过深层神经网络实现了难以置信的准确性，这在以前是不可能的。例如，你与 Alexa、谷歌搜索和谷歌照片的互动都是基于深度学习的，我们越使用它们，它们就越准确。在医学领域，来自深度学习、图像分类和目标识别的人工智能技术，现在可以像训练有素的放射科医生一样，在核磁共振成像上准确地发现癌症。

人工智能得益于数据。当算法是自我学习时，数据本身可以成为知识产权，答案就在数据中。人们只需要应用人工智能就可以把它们计算出来。现在数据的作用比以往任何时候都重要，它可以创造竞争优势。如果你在竞争激烈的行业中拥有最好的数据，即使每个人都在应用类似的技术，谁拥有最好的数据，谁就会胜出。

# Text B

Artificial Intelligence is everywhere, from Apple's iPhone keyboard to Zillow's home price estimates. There's also a lot of stuff out there that marketers are calling AI, but really isn't.

Perhaps things reached a new high point last month when AlphaGo, a virtual player of the ancient Chinese board game Go developed byAlphabet's DeepMind AI research group, *trounced* the top human player in the world, China's Ke Jie.

A moment of drama *encapsulates* the achievement: After Jie *resigned* in the second of three matches, the 19-year-old lingered in his chair, staring down at the board for several minutes, *fidgeting*

| *New Words and Expressions* |
| --- |
| **trounce**/traʊns/vt. |
| 痛打；严责；打败 |
| **encapsulate** /ɪnˈkæpsjʊleɪt/ vt. |
| 将…封进内部；概述 |
| **resign** /rɪˈzaɪn/ vi. |
| 辞职；放弃 |
| **fidget**/ˈfɪdʒɪt/vi. |
| 烦躁；坐立不安 |

with game pieces and *scratching* his head. Aja Huang, the DeepMind senior research scientist who was tasked with moving game pieces on behalf of AlphaGo, eventually got up from his chair and walked offstage, leaving Jie alone for a moment.

Still, it's generally true that a human being like Jie has more brainpower than a computer. That's because a person can perform a wide range of tasks better than machines, while a given computer program enhanced with AI like AlphaGo might be able to *edge out* a person at just a few things.

But the prospect of AI becoming smarter than people at most tasks is the single biggest thing that drives debates about effects on employment, creativity and even human existence.

Here's an overview of what AI really is, and what the biggest companies are doing with it.

So what is AI, really?

Given that everybody's talking about AI now, you would think it's new. But the underlying techniques are not. The field got its start in the mid-twentieth century, and one of its most popular methods came about in the 1980s.

AI first took hold in the 1950s. While some of its underlying concepts predate it, the term itself dates to 1956, when John McCarthy, a math professor at Dartmouth College, proposed a summer research project based on the idea that "every aspect of learning or any other feature of intelligence can in principle be so precisely described that a machine can be made to *simulate* it."

In the next few years AI research labs popped up at the Massachusetts Institute of Technology (MIT) and Stanford University. Research touched on computer chess, robotics and natural-language communication.

Interest in the field *fluctuated* over time. AI winters came in the 1970s and 1980s as public interest *waned* and outside funding dried up. Startups boasting promising capabilities and venture capital backing in the mid-1980s abruptly disappeared, as John Markoff detailed in his 2015 book "Machines of Loving Grace."

There are several other terms you often hear in connection to AI.

Machine learning generally *entails* teaching a machine how to do a particular thing, like recognizing a number, by feeding it a bunch of data and then directing it to make predictions on new data.

**New Words and Expressions**

**scratch** /skrætʃ/ vt.
抓；刮

**edge out**
替代；微微胜过

**simulate**/ˈsɪmjʊleɪt/ vt.
模仿

**fluctuate**/ˈflʌktʃueɪt/ vt.
使波动；使动摇

**wane**/weɪn/ vi.
衰落；变小

**entail**/ɛnˈteɪl/ vt.
使需要，必需；承担

The big deal about machine learning now is that it's getting easier to invent software that can learn over time and get smarter as it accumulates more and more data. Machine learning often requires people to hand-engineer certain features for the machine to look for, which can be complex and time-consuming.

Deep learning is one type of machine learning that demands less hand-engineering of features. Often the approach involves artificial neural networks, a mathematical system loosely inspired by the way neurons work together in the human brain. Neuroscientist Warren McCulloch and mathematician Walter Pitts came up with the first such system in 1943. Through the years, researchers advanced the concept with various techniques, including adding multiple layers. With each *successive* layer, higher-level features could be detected in the original data to make a better prediction. The layers pick out features in the data themselves. But using more layers demands more computing power.

Why is it suddenly so hot?

Through the years, hardware has gotten more powerful, and chipmakers including Nvidia have refined their products to better suit AI computations. Larger data sets in many domains have become available to train models more *extensively* as well.

In 2012, Google made headlines when it trained a neural network with 16,000 central processing unit (CPU) chips on 10 million images from YouTube videos and taught it to recognize cats. But later that year, the world of image recognition was rocked when an eight-layer neural network trained on two graphics processing units (GPUs) outdid all others in a competition to accurately classify images based on their content. Months later, Google acquired DNNresearch, the University of Toronto team behind the breakthrough.

Since then, AI activity has only accelerated, with the world's top technology companies leading the way.

Meanwhile, the world's most valuable companies — technology companies! — continue to publish research on their latest gains, which only adds to the *fascination*.

Google and its parent company Alphabet have made several AI Beyond that, perhaps in a few decades, an AI system with superhuman capabilities in most domains — sometimes referred to as artificial general intelligence — could emerge. Depending on whom you ask, that could be either very good or very bad. In an extreme

**New Words and Expressions**

**successive** /sək'sesɪv/ adj.
连续的；继承的；依次的

**extensively** /ik'stensivli/ adv.
广阔地；广大地

**fascination** /fæsɪ'neɪʃ(ə)n/ n.
魅力；魔力；入迷

case, an AGI system could end up making humans *extinct*. But if things go well, perhaps AGI will be something that will *supercharge* humans and help them live longer. The prospect of either of these two *scenarios* is perhaps what draws so much attention to AI development today, and what has inspired so much science fiction in the past.

But for now, what people generally see is narrow AI — intelligence applied to a small number of domains — and it doesn't always work the way it should. Look at Alexa, Cortana, the Google Assistant or Siri — they misunderstand spoken words all the time.

The thing is, the biggest companies in the world are now investing in AI like never before. And that trend is not about to let up.

> **New Words and Expressions**
> extinct/ɪkˈstɪŋkt/ adj.
> 灭绝的，绝种的；熄灭的
> supercharge/ˈsuːpətʃɑːdʒ/ vt.
> 对…增压；使…超负荷
> scenarios/sɪˈnɛrɪəuz/ n.
> 情节；脚本

## 参考译文 B

从苹果的 iPhone 键盘到 Zillow 的房价估算，人工智能无处不在。市场上也有很多营销人员称之为人工智能的东西，但实际上并非如此。

上个月，由 Alphabet 旗下 DeepMind AI 研究集团开发的围棋虚拟玩家 AlphaGo 击败了来自中国的世界顶级棋手柯洁，让人工智能达到了一个新的高潮。

在这背后有一个戏剧性的时刻：在三场比赛的第二场比赛中，19 岁的柯洁几乎崩溃。他坐在椅子上，低头盯着棋盘看了几分钟，一边摆弄棋子，一边挠头。负责替 AlphaGo 移动棋子的 DeepMind 高级研究科学家阿贾·黄最终从椅子上站起来，走下赛场，让柯洁独自待了一会儿。

尽管如此，像柯洁这样的人确实比计算机更聪明。因为人类可以比机器更好地执行各种各样的任务，而具有人工智能的计算机程序，例如 AlphaGo，只能在某些事情上胜过人类。

但是，在一些任务中人工智能表现得比人类聪明，因此引发了关于就业、创造力乃至人类生存影响的争论。

以下是关于人工智能的概述，以及大公司正在用它做什么。

那么人工智能到底是什么呢？

因为现在每个人都在谈论人工智能，你可能会认为它是新的。但它的基本技术并不新。这个领域在 20 世纪中期开始发展，最流行的方法之一出现在 20 世纪 80 年代。

人工智能的首次提出是在 20 世纪 50 年代。虽然它的一些基础概念在此之前就已经出现，但这个词本身可以追溯到 1956 年，由达特茅斯学院的数学教授约翰·麦卡锡（John McCarthy）在一个暑期研究项目中提出，该项目基于这样一个理念：学习的方方面面以及智能的其他特征都可以精确描述，从而能够进行机器仿真。

在接下来的几年里，麻省理工学院（MIT）和斯坦福大学（Stanford University）成立了人工智能研究实验室，研究涉及计算机象棋、机器人和自然语言交流。

人们对这一领域的兴趣随着时间而发生了变化。随着公众兴趣的减退和外部资金的枯竭，20世纪70年代和80年代成了人工智能的寒冬。正如约翰·马尔科夫（John Markoff）在2015年出版的 *Machines of Loving Grace* 一书中所描述的那样，在20世纪80年代中期，那些拥有良好能力和风险资本支持的初创企业突然消失了。

下面是几个人们经常听到的与人工智能有关的术语。

机器学习通常需要教会机器如何做一件特定的事情，例如识别一个数字，给它输入一组数据，然后引导它对新数据进行预测。

现在机器学习的重要之处在于，随着时间的推移，积累越来越多的数据，开发出能够学习，以及变得更智能的软件，这变得越来越容易。机器学习通常需要人们手工设计某些特性，以便机器寻找，这可能是复杂和耗时的。

深度学习是一种机器学习，它对手动操作要求较少。这种方法通常涉及人工神经网络，这是一种数学系统，其灵感来自于人类大脑中神经元协同工作的方式，神经学家沃伦·麦卡洛克和数学家沃尔特·皮茨在1943年第一次提出。多年来，研究人员通过各种技术发展了这一概念，其中包括多层感知技术。对于每一个连续的层，可以在原始数据中检测到更高层次的特征，从而做出更好地预测。这些层在数据中挑选，但是使用更多的层需要更强的计算能力。

人工智能为什么突然备受关注？

这些年，硬件变得越来越强大，包括英伟达（Nvidia）在内的芯片制造商已经改进了产品，以更好地适应人工智能计算。在许多领域中，更大的数据集也可用于广泛的训练模型。

2012年，谷歌用16,000个中央处理器（CPU）芯片对YouTube视频中的1000万张图片进行神经网络训练，并教会它识别猫，这让它成为头条新闻。但同年晚些时候，在一场根据图像内容进行精确分类的竞赛中，使用两个图形单元（GPU）训练的八层神经网络击败了其他对手，震惊了图像识别领域。几个月后，谷歌收购了取得这一突破性进展的多伦多大学DNNresearch团队。

从那以后，人工智能发展更为迅速，全球顶尖科技公司走在了前面。同时，世界上最有价值的科技公司继续发表它们的最新研究成果，这让人对人工智能更加着迷。

谷歌及其母公司Alphabet已经开发了数个人工智能系统，或许在几十年内，一个在大多数领域都具备超人能力的人工智能系统将会出现。这可能是好的，也可能是坏的。在极端情况下，AGI系统可能最终导致人类灭绝。但如果一切进展顺利，AGI系统或许能使人的能力更强，或帮助人们活得更久。这两种情况中任何一种，都引起了人们对人工智能的关注，也赋予了科幻小说许多灵感。

但就目前而言，人们通常看到的是狭隘的人工智能，应用于少数领域的智能，它并不总是以它应该的方式工作。看看Alexa、Cortana、谷歌助理或Siri，它们总是误解别人说的话。

世界上的许多大公司正在前所未有地投资人工智能，这种趋势不会停止。

# Chapter *2*

# History of Artificial Intelligence

## Text A

The history of Artificial Intelligence (AI) began in *antiquity* (Figure 2-1), with myths, stories and *rumors* of artificial beings endowed with intelligence or consciousness by master *craftsmen*; as Pamela McCorduck writes, AI began with "an ancient wish to *forge* the gods".

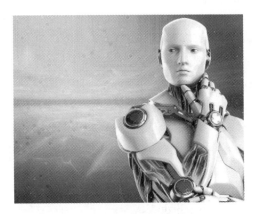

Figure 2-1　Artificial Intelligence

The study of mechanical or "formal" reasoning began with philosophers and mathematicians in antiquity. The study of mathematical logic led directly to Alan Turing's theory of computation, which suggested that a machine, by *shuffle* symbols as simple as "0" and "1", could *simulate* any *conceivable* act of mathematical *deduction* (Figure 2-2). This insight, that digital computers can simulate any

**antiquity**/ænˈtɪkwɪtɪ/ n.

古物；古代

**rumor**/ˈruːmə/ n.

谣言；传闻

**craftsman**/ˈkrɑːf(t)smən/ n.

工匠；手艺人

**forge**/fɔːdʒ/ v.

伪造；锻造

**shuffle**/ˈʃʌfl/ v.

把...变换位置；调动

**simulate** /ˈsɪmjʊleɪt/ v.

模仿；假装

**conceivable** /kənˈsiːvəb(ə)l/ adj.

可能的；可想到的

**deduction** /dɪˈdʌkʃ(ə)n/ n.

扣除，减除；推论

process of formal reasoning, is known as the Church-Turing thesis[1]. Along with concurrent discoveries in *neurobiology*, information theory and *cybernetics*, this led researchers to consider the possibility of building an electronic brain. Turing proposed that "if a human could not distinguish between responses from a machine and a human, the machine could be considered "intelligent". The first work that is now generally recognized as AI was McCullouch and Pitts' 1943 formal design for Turing-complete "artificial neurons".

***New Words and Expressions***

**neurobiology**/ˌnjʊərəʊbaɪˈɒlədʒɪ/ n.
神经生物学

**cybernetics**/saɪbəˈnetɪks/ n.
控制论

**checkers**/ˈtʃɛkəz/ n.
跳棋

**strategy**/ˈstrætədʒɪ/ n.
战略，策略

**algebra**/ˈældʒɪbrə/ n.
代数学

**substantially**/səbˈstænʃ(ə)lɪ/ adv.
相当多地；实质上

**ongoing** /ˈɒngəʊɪŋ/ adj.
不间断的，进行的；前进的

Figure 2-2    Turing Test

The field of AI research was born at a workshop at Dartmouth College in 1956. Attendees Allen Newell (CMU), Herbert Simon (CMU), John McCarthy (MIT) (Figure 2-3), Marvin Minsky (MIT) and Arthur Samuel (IBM) became the founders and leaders of AI research. They and their students produced programs that the press described as "astonishing": computers were learning *checkers strategies* (c. 1954 and by 1959 were reportedly playing better than the average human), solving word problems in *algebra*, proving logical theorems (Logic Theorist, first run c. 1956) and speaking English. By the middle of the 1960s, research in the U.S. was heavily funded by the Department of Defense and laboratories had been established around the world. AI's founders were optimistic about the future: Herbert Simon predicted, "machines will be capable, within twenty years, of doing any work a man can do". Marvin Minsky agreed, writing, "within a generation ... the problem of creating 'artificial intelligence' will *substantially* be solved".

They failed to recognize the difficulty of some of the remaining tasks. Progress slowed and in 1974, in response to the criticism of Sir James Lighthill and *ongoing* pressure from the U.S. Congress to fund more productive projects, both the U.S. and British governments cut off exploratory research in AI. The next

Figure 2-3　John McCarthy

*New Words and Expressions*

**analytical skills**
分析技巧

**restore**/rɪˈstɔː/ v.
恢复；修复

**collapse** /kəˈlæps/ v.
倒塌；瓦解

**disrepute**/ˌdɪsrɪˈpjuːt/ n.
不光彩，坏名声

**hiatus**/haɪˈeɪtəs/ n.
裂缝，空隙

**data mining**
数据挖掘技术

**medical diagnosis**
医疗诊断

**reign**/reɪn/ v.
统治；支配

**quiz show**
智力竞争节目

**margin**/ˈmɑːdʒɪn/ n.
边缘；利润

**perception**/pəˈsepʃ(ə)n/ n.
知觉；感觉

**benchmark**/ˈben(t)ʃmɑːk/ n.
基准；标准检查程序

few years would later be called an "AI winter", a period when obtaining funding for AI projects was difficult.

In the early 1980s, AI research was revived by the commercial success of expert systems, a form of AI program that simulated the knowledge and *analytical skills* of human experts. By 1985, the market for AI had reached over a billion dollars. At the same time, Japan's fifth generation computer project inspired the U.S. and British governments to *restore* funding for academic research. However, beginning with the *collapse* of the Lisp Machine market in 1987, AI once again fell into *disrepute*, and a second, longer-lasting *hiatus* began.

In the late 1990s and early 21st century, AI began to be used for logistics, *data mining*, *medical diagnosis* and other areas. The success was due to increasing computational power (see Moore's law[2]), greater emphasis on solving specific problems, new ties between AI and other fields (such as statistics, economics and mathematics), and a commitment by researchers to mathematical methods and scientific standards. Deep Blue[3] became the first computer chess-playing system to beat a *reigning* world chess champion, Garry Kasparov, on 11 May 1997.

In 2011, a Jeopardy! *quiz show* exhibition match, IBM's question answering system, Watson, defeated the two greatest Jeopardy! champions, Brad Rutter and Ken Jennings, by a significant *margin*. Faster computers, algorithmic improvements, and access to large amounts of data enabled advances in machine learning and *perception*; data-hungry deep learning methods started to dominate accuracy *benchmarks* around 2012. The Kinect, which provides a 3D body–motion interface for the Xbox 360 and the Xbox One, uses algorithms that emerged from lengthy AI research as do intelligent personal assistants in smartphones. In March 2016, AlphaGo won 4 out of 5 games of Go in a match with Go champion Lee Sedol, becoming the first computer Go-playing system to beat a professional Go player without *handicaps*. In the 2017 Future of Go Summit, AlphaGo won a three-game match with Ke Jie, who at the time continuously held the world No. 1 ranking for two years. This marked the completion of a significant *milestone* in the development of Artificial Intelligence as Go is an extremely complex game, more so than Chess.

According to Bloomberg's Jack Clark, 2015 was a landmark year for artificial intelligence, with the number of software projects that use AI within Google increased from a "sporadic usage" in 2012 to more than 2,700 projects. Clark also presents factual data indicating that error rates in image processing tasks have fallen significantly since 2011. He *attributes* this to an increase in affordable *neural networks*, due to a rise in cloud computing infrastructure and to an increase in research tools and datasets. Other cited examples include Microsoft's development of a Skype system that can *automatically* translate from one language to another and Facebook's system that can describe images to blind people. In a 2017 survey, one in five companies reported they had "incorporated AI in some offerings or processes". Around 2016, China greatly accelerated its government funding; given its large supply of data and its rapidly increasing research output, some observers believe it may be on track to becoming an "AI superpower" (Figure 2-4).

***New Words and Expressions***

**handicap**/ˈhændɪkæp/ n.
　障碍；不利条件

**milestone**/ˈmaɪlstəʊn/ n.
　里程碑，划时代的事件

**attribute**/əˈtrɪbjuːt/ v.
　归属；把…归于

**neural network**
　神经网络

**automatically**/ɔːtəˈmætɪklɪ/ adv.
　自动地；机械地

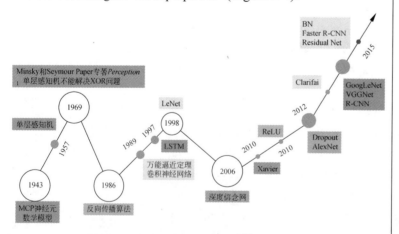

Figure 2-4　Timeline of Artificial Intelligence

## Terms

### 1. Church-Turing thesis

丘奇-图灵论题（The Church-Turing thesis）是计算机科学中以数学家阿隆佐·丘奇(Alonzo Church)和阿兰·图灵命名的论题。该论题最基本的观点表明，所有计算或算法都可以由一台图灵机来执行。以任何常规编程语言编写的计算机程序都可以翻译成一台图灵机，反之任何一台图灵机也都可以翻译成大部分编程语言大程序，所以该论题和以下说法等价：常规的编程语言可以足够有效地表达任何算法。该论题被普遍假定为真，也被称为丘奇论题或丘奇猜想和图灵论题。

### 2. Moore's law

摩尔定律（Moore's law）是由英特尔（Intel）创始人之一戈登·摩尔（Gordon Moore）提出来的。其内容是：当价格不变时，集成电路上可容纳的元器件的数目，每隔 18～24 个月便会增加一倍，性能也将提升一倍。换言之，每一美元所能买到的计算机性能，将每隔 18～24 个月翻一倍以上。这一定律揭示了信息技术进步的速度。

### 3. Deep Blue

"深蓝（Deep Blue）"是美国 IBM 公司生产的一台超级国际象棋计算机，重 1270 千克，有 32 个大脑（微处理器），每秒钟可以计算 2 亿步。"深蓝"输入了一百多年来优秀棋手的对局共计两百多万局。

## Comprehension

### Blank Filling

1. Artificial intelligence was first proposed in _____.

2. Alan Turing's theory suggested that a machine, by shuffling symbols as simple as "0" and "1", could simulate any conceivable act of _____.

3. Along with concurrent discoveries in _____, _____ and _____, this led researchers to consider the possibility of building an electronic brain.

4. _____ is recognized as the first complete artificial intelligence works with Turing.

5. In the early 1980s, AI research was revived by the commercial success of _____.

### Content Questions

1. In the late 1990s and early 21st century, where is artificial intelligence starting to be used?

2. What makes artificial intelligence successful?

3. What advances have been made in machine learning and perception?

4. Clark thinks what is the reason for the significant decrease in the error rates in image processing tasks?

## Answers

### Blank Filling

1. 1956
2. mathematical deduction
3. neurobiology; information theory; cybernetics
4. Artificial neurons
5. expert systems

### Content Questions

1. AI began to be used for logistics, data mining, medical diagnosis and other areas in the late 1990s and early 21st century.
2. The success was due to increasing computational power, greater emphasis on solving specific problems, new ties between AI and other fields and a commitment by researchers to mathematical methods and scientific standards.
3. Faster computers, algorithmic improvements, and access to large amounts of data enabled advances in machine learning and perception.
4. He attributes this to an increase in affordable neural networks, due to a rise in cloud computing infrastructure and to an increase in research tools and datasets.

## 参考译文 A

人工智能（AI）的历史可以追溯到远古时代，当时的神话、故事和传说都是由工匠大师赋予智能或意识；正如帕梅拉·麦考达克（Pamela McCorduck）所写，人工智能始于"铸造神灵的古老愿望"（图 2-1）。

对机械推理或"形式"推理的研究始于古代的哲学家和数学家。艾伦·图灵（Alan Turing）的计算理论就是基于这些数理逻辑的研究，该理论认为，一台机器通过对"0"和"1"这样简单的符号进行变换，能够模拟任何可以想象的数学推导行为（图 2-2）。数字计算机可以模拟形式推理的任何过程，这种观点被称为丘奇-图灵论题。神经生物学、信息论和控制论的同时发现促使研究人员思考构建一个电子大脑的可能性。图灵提出，如果无法区分回答来自机器还是人，那么这台机器可以被认为是"智能的"。现在公认的第一个人工智能作品是麦卡洛克和皮茨于 1941 年设计的"人工神经元"模型。

人工智能研究领域诞生于 1956 年达特茅斯学院的一个研讨会上。参会人 Allen Newell（CMU）、Herbert Simon（CMU）、John McCarthy（MIT）（图 2-3）、Marvin Minsky（MIT）和 Arthur Samuel（IBM）成为人工智能研究的创始人和领导者。他们和学生制作了一系列被媒体描述为"惊人"的程序：计算机正在学习跳棋策略（据报道，到 1959 年，计算机的表现已经超过了普通人），正在解决代数中的单词问题，正在证明逻辑定理（逻辑理论家，大约 1956 年首次运行），正在说英语。到 20 世纪 60 年代中期，美国的研究得到了国防部的大量资助，世界各地都建立了实验室。人工智能的创始人对未来持乐观态度：赫伯特·西蒙（Herbert Simon）预言"机器将在 20 年内完成人类能做的任何工作。"马文·明斯基（Marvin Minsky）对此表示赞同，他写道："一代人之内……创造'人工智能'的问题将得到实质性解决"。

　　之前过于乐观，他们并没有认识到还有很多困难需要克服。由于詹姆斯·莱特希尔爵士（Sir James Lighthill）的批评，以及美国国会要求资助更有成效的项目的持续压力，美国和英国政府在 1974 年都停止了人工智能领域的探索性研究。接下来的几年被称为"人工智能的冬天"，这段时间人工智能项目很难获得资金。

　　20 世纪 80 年代初，专家系统的商业成功使人工智能恢复生机。专家系统是一种模拟人类专家知识和分析技能的人工智能程序。到 1985 年，人工智能的市场已经超过了 10 亿美元。与此同时，日本的第五代计算机项目激发了美国和英国政府，他们恢复对学术研究的资助。然而，从 1987 年 Lisp 机器市场崩溃开始，人工智能又一次声名狼藉，第二次更持久的停滞开始了。

　　20 世纪 90 年代末和 21 世纪初，人工智能开始被用于物流、数据挖掘、医疗诊断等领域。人工智能的成功得益于计算能力的增强（如摩尔定律）、对解决特定问题的重视程度的提高、人工智能与其他领域（如统计、经济学和数学）之间的新联系，以及研究人员对数学方法和科学标准的投入。1997 年 5 月 11 日，"深蓝"成为首个击败国际象棋世界冠军加里·卡斯帕罗夫的计算机国际象棋系统。

　　2011 年，在一个名为 *Jeopardy!* 的智力竞赛节目中，IBM 的问答系统——沃森以显著的优势击败了该节目两个最伟大的冠军布拉德·拉特（Brad Rutter）和肯·詹宁斯（Ken Jennings）。计算机速度的增强，算法的改进，以及对大量数据的访问，使机器学习和感知的进步成为可能；大流量数据的深度学习方法在 2012 年左右开始控制精度基准。Kinect 为 Xbox 360 和 Xbox One 提供了一个 3D 身体运动界面，它使用的算法和智能手机中的智能助手一样，都是经过长期人工智能研究得出的。2016 年 3 月，AlphaGo 在与围棋冠军李世石的对弈中，五局四胜，成为首个击败无残障职业棋手的计算机围棋系统。在 2017 年的未来围棋峰会上，AlphaGo 以 3:0 赢得了连续两年蝉联世界第一的柯洁。这是人工智能发展的一个重要里程碑，因为围棋是一项比国际象棋更为复杂的游戏。

　　布隆伯格（Bloomberg）公司的杰克·克拉克（Jack Clark）表示，2015 年是人工智能具有里程碑意义的一年，在谷歌中使用人工智能的软件项目从 2012 年的"零星使用"增加到 2700 多个。克拉克还提供了实际数据，表明自 2011 年以来，图像处理任务中的错误率已经显著下降。他将此归因于神经网络的发展，云计算基础设施以及研究工具和数据集的增加。其他例子还包括微软开发的 Skype 系统可以自动从一种语言转换到另一种语言，以及 Facebook 的系统可以向盲人描述图像。在 2017 年的一项调查中，五分之一的公司表示，它们"在一些产品或流程中加入了人工智能"。2016 年前后，中国大幅加快政府资金投入；鉴于其庞大的数据供应和快速增长的研究产出，一些观察人士认为，中国可能正走上成为"人工智能超级大国"的道路（图 2-4）。

# Text B

## Complex AI Systems Explain Their Actions

In the future, service robots *equipped* with artificial intelligence (AI) are bound to be a common sight. These bots will help people navigate crowded airports, serve meals, or even schedule meetings.

Figure 2-5　CoBot

As these AI systems become more integrated into daily life, it is vital to find an efficient way to communicate with them. It is obviously more natural for a human to speak in *plain* language rather than a string of code. Further, as the relationship between humans and robots grows, it will be necessary to engage in conversations, rather than just give orders.

This *human-robot interaction* is what Manuela M. Veloso's research is all about. Veloso, a professor at Carnegie Mellon University, has focused her research on CoBots (Figure 2-5), *autonomous* indoor mobile service robots which transport items, guide visitors to building locations, and traverse the halls and elevators. The CoBot robots have been successfully autonomously navigating for several years now, and have traveled more than 1,000km. These accomplishments have enabled the research team to pursue a new direction, focusing now on novel human-robot interaction.

"If you really want these autonomous robots to be in the presence of humans and interacting with humans, and being capable of benefiting humans, they need to be able to talk with humans" Veloso says.

*New Words and Expressions*

**equip**/ɪˈkwɪp/ v.
装备，配备

**plain**/pleɪn/ adj.
平的；简单的

**human-robot interaction**
人机交互

**autonomous**/ɔːˈtɒnəməs/ adj.
自治的；自主的

## Communicating With CoBots

Veloso's CoBots are capable of autonomous localization and navigation in the Gates-Hillman Center using WiFi, LIDAR, and/or a Kinect *sensor* (yes, the same type used for video games).

The robots navigate by *detecting* walls as planes, which they match to the known maps of the building. Other objects, including people, are detected as *obstacles*, so navigation is safe and *robust*. Overall, the CoBots are good navigators and are quite consistent in their motion. In fact, the team noticed the robots could wear down the *carpet* as they traveled the same path numerous times.

Because the robots are autonomous, and therefore capable of making their own decisions, they are out of sight for large amounts of time while they navigate the multi-floor buildings. The research team began to wonder about this unaccounted time. How were the robots perceiving the environment and reaching their goals? How was the trip? What did they plan to do next?

"In the future, I think that *incrementally* we may want to query these systems on why they made some choices or why they are making some recommendations," explains Veloso.

The research team is currently working on the question of why the CoBots took the route they did while autonomous. The team wanted to give the robots the ability to record their experiences and then transform the data about their routes into natural language. In this way, the bots could communicate with humans and reveal their choices and hopefully the rationale behind their decisions.

## Levels of Explanation

The "internals" underlying the functions of any autonomous robots are completely based on *numerical* computations, and not natural language. For example, the CoBot robots in particular compute the distance to walls, assigning *velocities* to their motors to enable the motion to specific map *coordinates*.

Asking an autonomous robot for a *non-numerical* explanation is complex, says Veloso. Furthermore, the answer can be provided in many potential levels of detail.

"We define what we call the '*verbalization* space' in which this translation into language can happen with different levels of detail, with different levels of locality, with different levels of specificity."

---

### New Words and Expressions

**sensor**/ˈsensə/ n.
传感器

**detect**/dɪˈtekt/ v.
察觉；发现

**obstacle**/ˈɒbstək(ə)l/ n.
障碍，干扰；妨害物

**robust**/rə(ʊ)ˈbʌst/ adj.
强健的；健康的

**carpet**/ˈkɑːpɪt/ n.
地毯

**incrementally**/ˌɪnkrɪˈmentəlɪ/ adj.
增加的，增值的

**numerical**/njuːˈmerɪk(ə)l/ adj.
数值的；数字的

**velocity** /vəˈlɒsəti/ n.
【物】速度

**coordinate**/kəʊˈɔːdɪneɪt/ n.
坐标；同等的人或物

**non-numerical** n.
非数值

**verbalization**/ˈvɜːbəlaɪˌzeɪʃən/ n.
以言语表现；冗长；变成动词

For example, if a developer is asking a robot to detail their journey, they might expect a lengthy retelling, with details that include battery levels. But a random visitor might just want to know how long it takes to get from one office to another.

Therefore, the research is not just about the translation from data to language, but also the acknowledgment that the robots need to explain things with more or less detail. If a human were to ask for more detail, the request *triggers* CoBot "to move" into a more detailed point in the verbalization space.

"We are trying to understand how to empower the robots to be more trustable through these explanations, as they attend to what the humans want to know," says Veloso. The ability to generate explanations, in particular at multiple levels of detail, will be especially important in the future, as the AI systems will work with more complex decisions. Humans could have a more difficult time inferring the AI's reasoning. Therefore, the bot will need to be more transparent.

For example, if you go to a doctor's office and the AI there makes a recommendation about your health, you may want to know why it came to this decision, or why it recommended one medication over another.

Currently, Veloso's research focuses on getting the robots to generate these explanations in plain language. The next step will be to have the robots incorporate natural language when humans provide them with feedback. "[The CoBot] could say, 'I came from that way,' and you could say, 'well next time, please come through the other way,'" explains Veloso.

These sorts of corrections could be programmed into the code, but Veloso believes that "trustability" in AI systems will benefit from our ability to *dialogue*, *query*, and correct their autonomy. She and her team aim at contributing to a multi-robot, multi-human *symbiotic* relationship, in which robots and humans coordinate and cooperate as a function of their limitations and strengths.

"What we're working on is to really empower people — a random person who meets a robot — to still be able to ask things about the robot in natural language," she says.

In the future, when we will have more and more AI systems that are able to perceive the world, make decisions, and support

**New Words and Expressions**

**trigger**/ˈtrɪgə/ n.
触发器
**dialogue** /ˈdaɪəlɒg/ v.
对话
**query**/ˈkwɪərɪ/ v.
询问；对…表示疑问
**symbiotic**/sɪmbaɪˈɒtɪk/ adj.
共生的；共栖的

human decision-making, the ability to engage in these types of conversations will be essential.

This article is part of a Future of Life series on the AI safety research grants, which were funded by generous donations from Elon Musk and the Open Philanthropy Project.

## 参考译文 B

**复杂人工智能系统能解释他们的行为**

在未来几年，配备人工智能的服务型机器人将成为普遍趋势。这些机器人将会帮助人们在拥挤的机场导航，服务人们用餐，或者制订会议日程。

当人工智能系统变成人们日常生活中不可分割的一部分时，找到一个和机器有效的交流方式变得至关重要。对于人类来说，使用普通语言交流显然比用一串串代码沟通更为自然。与此同时，随着人类和机器的关系变得日益亲密，相较人类单方面的给出指令，双方皆能参与到对话变得尤为必要。

这种人机互动正是曼纽拉·M.维罗索（Manuela M. Veloso）教授的科研方向。维罗索是卡耐基·梅隆大学计算机科学系教授，她的研究集中于协作机器人。这些服务型机器人可以在室内运输物品，也可以为游客在楼宇间及楼内进行导航。这类协作机器人已经成功提供自动导航服务数年，累计超过 1000 千米。在这些成就的基础上，研究团队也在寻求新的研究方向，他们现在专注于新型人机交互。

"如果你真希望这类自动化机器人呈现于世人面前，并为人类生活提供便利，那么它们就需要能够和人类对话沟通。"维罗索说。

**与协作机器人沟通**

维罗索的协作机器人可以利用无线网络、激光雷达和体感装置（是的，就是用于游戏中的那种）在盖茨-希尔曼中心自主定位并导航。

机器人通过探测墙壁来导航，通过这种方式以匹配建筑的平面图。包括人类在内的其他物体则被定义为障碍物，所以导航系统是非常安全有效的。 整体来说，协作机器人是很好的导航器，并且在工作状态下始终如一。事实上，研究团队发现机器人会重复选择同一条路线，如此多次之后途经的地毯竟出现了磨损的迹象。

这是因为机器人的行为是自主的，它们有能力做出自己的决定。这样一来它们穿梭于各个楼层之间的大部分时间都在人类的视野范围之外。研究团队对这段行踪不明的时间非常好奇——机器人是如何感知周边环境并最终到达预定目标地点的？这些行程怎么样？它们的下一步行动计划是什么？

"在未来，我们会逐渐了解到这些系统是如何做选择，以及如何提出建议的。"维罗索说。

目前研究团队正致力于解释协作机器人在自主行动时是如何选定特定行动路线的。研究团队想赋予机器人记录能力，这样它们就可以把路线数据转化成自然语言。通过这种方法，机器人便能和人类进行沟通并给出他们的选择，并解释他们做出各种选择的基本原理。

### 不同等级的解释

任何自主机器人的内部分析功能都是依靠数值计算，而不是自然语言。举例来说，协作机器人根据距离墙的远近的估算来设定发动机速度以到达特定的位置。

维罗索认为让自主机器人做出非数值解释是非常复杂的。此外，机器人提供的答案里包含着不同的细节。"机器人给出的解释转化成语言将呈现出不同程度的细节，不同具体度的位置，以及不同级别的特征。我们把这种情况定义为'语言化空间'。"

例如，如果开发者要求机器人提供所走线路的细节数据，他们想知道的是路程长度外加电池消耗等细节。但是一个随机访问用户也许只是想知道机器人从一个办公室到另一个办公室花了多长时间而已。

因此，研究人员不只是单纯地把数据转化成语言，而是让机器人给出解释时有能力界定提供数据的细节程度。如果询问者想要请求更多的细节，那么该请求就会触发机器人'语言化空间'中更加细节的部分。

维罗索说："我们正在试图了解如何让机器人在回答此类问题时有更强的自主能力，让它们理解人类到底想获取什么信息。"在今后，人工智能系统会参与到更加复杂的决策中，所以给出不同细节的解释的能力将会变得尤为重要。人类推断人工智能推理过程时也会变得更加复杂。因此，机器系统需要变得更加通透易懂。

例如，你去看医生时人工智能为你提出健康建议，你也许会想知道他们是基于什么得出这些结论的，或者为什么他们推荐这类药而不是其他药。

目前，维罗索的研究专注于如何让机器人产生简单易懂的语言。下一步将会是在人类给出反馈时让机器人理解人类自然语言。如果，协作机器人说："我是从这条路来的。"那你可以回答他："下一次请你使用其他的路径。"维罗索说。

这种行为矫正可以通过编程的方式写进系统，但是维罗索相信人工智能系统中的'信任度'可以通过与人类的对话，被询问和矫正来增强。她和她的团队致力于建立一个多机器人、多人类的共生关系。在这个体系中，机器人和人类互相协作，取长补短。

"我们现在正在努力实现任何使用者都可以用自然语言对机器人进行询问。"她说道。在未来，我们将拥有更多的人工智能系统来认识这个世界，做出决定，并辅助人类做出决定。那时，与人类对话的能力将会变得至关重要。

# Chapter *3*

# Machine Learning

## Text A

The name machine learning was coined in 1959 by Arthur Samuel. Machine learning (ML) is the scientific study of algorithms and *statistical* models that computer systems use to progressively improve their performance on a specific task. Machine learning algorithms build a mathematical model of *sample data*, known as "training data", in order to make predictions or decisions without being *explicitly* programmed to perform the task. Machine learning algorithms are used in the applications of email *filtering*, detection of network *intruders*, and computer vision, where it is infeasible to develop an algorithm of specific instructions for performing the task. Machine learning is closely related to computational statistics, which focuses on making predictions using computers. The study of *mathematical optimization* delivers methods, theory and application domains to the field of machine learning. Data *mining*[1] is a field of study within machine learning, and focuses on exploratory data analysis through unsupervised learning. In its application across business problems, machine learning is also *referred to as* predictive analytics.

### Machine Learning Tasks

A *support vector machine*[2] is a supervised learning model that divides the data into regions separated by a *linear boundary*. Here, the linear boundary divides the black circles from the white.

Machine learning tasks are classified into several broad *categories*. In supervised learning[3], the algorithm builds a mathematical model

| New Words and Expressions |
|---|
| **statistical model** 统计模型 |
| **sample data** 样本数据 |
| **explicitly**/ɪkˈsplɪsɪtlɪ; ek-/ adv. 明确地;清楚地 |
| **filter**/ˈfɪltə/ v. 过滤;渗透 |
| **intruder**/ɪnˈtruːdə/ n. 侵入者;干扰者 |
| **mathematical optimization** 数学最优化 |
| **mining**/ˈmaɪnɪŋ/ n. 矿业;采矿 |
| **referred to as** 被称为… |
| **support vector machine** 支持向量机 |
| **linear**/ˈlɪnɪə/ adj. 线的,线性的;直线的 |
| **boundary**/ˈbaʊnd(ə)rɪ/ n. 边界;范围 |
| **category**/ˈkætɪg(ə)rɪ/ n. 种类,分类 |

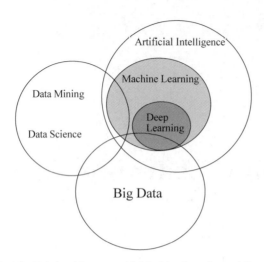

Figure 3-1    The Relationship among AI, Machine Learning and Deep Learning

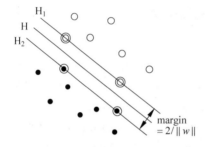

Figure 3-2    Optional Separating Hyperplane in Support Vector Machine

**New Words and Expressions**

**restrict**/rɪˈstrɪkt/ v.

　限制；约束

**feedback**/ˈfiːdbæk/ n.

　反馈；成果

**semi-supervised learning**

　半监督学习

**a portion of**

　一份

**folder**/ˈfəʊldə/ n.

　文件夹；折叠机

**spam**/spæm/ n.

　垃圾邮件

of a set of data that contains both the inputs and the desired outputs. For example, if the task were determining whether an image contained a certain object, the training data for a supervised learning algorithm would include images with and without that object (the input), and each image would have a label (the output) designating whether it contained the object. In special cases, the input may be only partially available, or *restricted* to special *feedback*. *Semi-supervised learning* algorithms develop mathematical models from incomplete training data, where *a portion of* the sample inputs are missing the desired output.

　　Classification algorithms and regression algorithms are types of supervised learning. Classification algorithms are used when the outputs are restricted to a limited set of values. For a classification algorithm that filters emails, the input would be an incoming email, and the output would be the name of the *folder* in which to file the email. For an algorithm that identifies *spam* emails, the output would be the prediction of either "spam" or "not spam", represented by the Boolean values one and zero. *Regression algorithms* are named for their continuous outputs, meaning they may have any value within a range. Examples of a continuous value are the temperature, length, or price of an object.

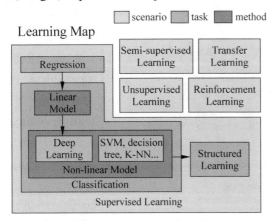

Figure 3-3   Machine Learning Tasks

In *unsupervised learning*, the algorithm builds a mathematical model of a set of data which contains only inputs and no desired outputs. Unsupervised learning algorithms are used to find structure in the data, like grouping or *clustering* of data points. Unsupervised learning can discover *patterns* in the data, and can group the inputs into categories, as in feature learning. *Dimensionality* reduction is the process of reducing the number of "features", or inputs, in a set of data.

Active learning algorithms access the desired outputs (training labels) for a limited set of inputs based on a budget, and *optimize* the choice of inputs for which it will acquire training labels. When used *interactively*, these can be presented to a human user for labeling. *Reinforcement learning* algorithms are given feedback in the form of positive or negative reinforcement in a *dynamic* environment, and are used in autonomous vehicles or in learning to play a game against a human *opponent*. Other specialized algorithms in machine learning include topic modeling, where the computer program is given a set of natural language *documents* and finds other documents that cover similar topics. Machine learning algorithms can be used to find the unobservable probability *density function* in *density estimation* problems. Meta learning algorithms learn their own *inductive bias* based on previous experience. In developmental robotics, robot learning algorithms generate their own *sequences* of learning

**New Words and Expressions**

**regression algorithm**
回归算法

**unsupervised learning**
无监督学习

**cluster**/ˈklʌstə/ v.
使聚集

**pattern**/ˈpæt(ə)n/ n.
模式；图案

**dimensionality**
/dɪˌmenʃəˈnælətɪ/ n.
维度

**optimize**/ˈɒptɪmaɪz/ v.
优化；持乐观态度

**interactive**/ɪntərˈæktɪv/ adj.
交互式的；相互作用的

**reinforcement learning**
强化学习

**dynamic**/daɪˈnæmɪk/ adj.
动态的；动力的

**opponent**/əˈpəʊnənt/ n.
对手；反对者

**document**/ˈdɒkjʊm(ə)nt/ n.
文档；证件

**density function**
密度函数

**density estimation**
密度估计

**inductive bias**
归纳偏向

**sequence**/ˈsiːkw(ə)ns/ n.
序列；顺序

experiences, also known as a *curriculum*, to *cumulatively* acquire new skills through *self-guided* exploration and social interaction with humans. These robots use guidance *mechanisms* such as active learning, *maturation*, motor synergies, and imitation.

Figure 3-4　Machine Learning

### Processes and Techniques

Various processes, techniques and methods can be applied to one or more types of machine learning algorithms to enhance their performance.

✓　Feature learning

Feature learning algorithms, also called representation learning algorithms, often attempt to preserve the information in their input but also transform it in a way that makes it useful, often as a *pre-processing* step before performing classification or predictions. This technique allows reconstruction of the inputs coming from the unknown data-generating distribution, while not being necessarily faithful to *configurations* that are *implausible* under that distribution. This replaces manual feature engineering, and allows a machine to both learn the features and use them to perform a specific task.

✓　*Sparse* dictionary learning

Sparse dictionary learning is a feature learning method where a training example is represented as a linear combination of basis functions, and is assumed to be a *sparse matrix*. The method is strongly *NP-hard*[4] and difficult to solve *approximately*. A popular heuristic method for sparse dictionary learning is the K-SVD algorithm[5]. Sparse dictionary learning has been applied in several contexts. In classification, the problem is to determine to which classes a previously unseen training example belongs. For a dictionary where each class has already been built, a new training example is associated with the class that is best sparsely represented by the

**New Words and Expressions**

**curriculum**/kəˈrɪkjʊləm/ n.
课程

**cumulatively**/ˈkjuːmjʊlətɪvlɪ/ adv.
累积地

**self-guided**/ˈself gaidid/ adj.
自导的；自动导向的

**mechanism**/ˈmek(ə)nɪz(ə)m/ n.
机制；原理

**maturation**/mætjʊˈreɪʃ(ə)n/ n.
成熟；化脓

**pre-processing**
预处理

**configuration**/kənˌfɪgəˈreɪʃ(ə)n/ n.
配置；结构

**implausible**/ɪmˈplɔːzɪb(ə)l/ adj.
难以置信的

**sparse**/spɑːs/ adj.
稀疏的；稀少的

**sparse matrix**
稀疏矩阵

**NP-hard**
NP 困难问题

**approximately**/əˈprɒksɪmətlɪ/ v.
近似；使…接近

corresponding dictionary. Sparse dictionary learning has also been applied in image *de-noising*. The key idea is that a clean image patch can be sparsely represented by an image dictionary, but the noise cannot.

✓ *Anomaly detection*

In data mining, anomaly detection, also known as outlier detection, is the identification of rare items, events or observations which raise *suspicions* by differing significantly from the majority of the data. Typically, the anomalous items represent an issue such as *bank fraud*, a structural defect, medical problems or errors in a text. Anomalies are referred to as *outliers*, *novelties*, noise, deviations and exceptions.

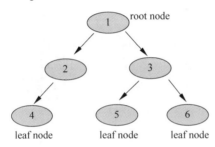

Figure 3-5   Schematic Diagram of Decision Tree

✓ *Decision trees*

Decision tree learning uses a decision tree as a predictive model to go from observations about an item (represented in the branches) to conclusions about the item's target value (represented in the leaves). It is one of the predictive modeling approaches used in statistics, data mining and machine learning. Tree models where the target variable can take a *discrete set* of values are called classification trees; in these tree structures, leaves represent class labels and branches represent conjunctions of features that lead to those class labels. Decision trees where the target variable can take continuous values (typically real numbers) are called regression trees. In decision analysis, a decision tree can be used to visually and explicitly represent decisions and decision making. In data mining, a decision tree describes data, but the resulting classification tree can be an input for decision making.

✓ *Association rules*

Association rule learning is a rule-based machine learning method for discovering relationships between variables in large

---

**New Words and Expressions**

**de-noising**
去噪
**anomaly detection**
异常检测
**suspicion**/sə'spɪʃ(ə)n/ n.
怀疑；嫌疑
**bank fraud**
银行欺诈
**outlier**/'aʊtlaɪə/ n.
异常值
**novelty**/'nɒv(ə)ltɪ/ n.
新奇；新奇的事物
**discrete set**
离散集

databases. It is intended to identify strong rules discovered in databases using some measure of "interestingness". This rule-based approach generates new rules as it analyzes more data. The ultimate goal, assuming the set of data is large enough, is to help a machine mimic the human brain's feature extraction and abstract association capabilities for data that has not been categorized.

## Terms

### 1. Data Mining

数据挖掘又译为资料探勘、数据采矿。它是数据库知识发现中的一个步骤。数据挖掘一般是指从大量的数据中通过算法搜索隐藏于其中信息的过程。数据挖掘通常与计算机科学有关，并通过统计、在线分析处理、情报检索、机器学习、专家系统（依靠过去的经验法则）和模式识别等诸多方法来实现上述目标。

### 2. Support Vector Machine

支持向量机（Support Vector Machine，SVM）是一类按监督学习方式对数据进行二元分类的广义线性分类器，在解决小样本、非线性及高维模式识别中表现出许多特有的优势，并能够推广应用到函数拟合等其他机器学习问题中。在机器学习中，支持向量机是与相关的学习算法有关的监督学习模型，可以分析数据、识别模式、用于分类和回归分析。

### 3. Supervised Learning

监督式学习是一种机器学习的方式，可以由训练资料中学到或建立一个模式，并依此模式推测新的实例。训练资料是由输入物件（通常是向量）和预期输出所组成。函数的输出可以是一个连续的值（称为回归分析），或是预测一个分类标签（称作分类）。

### 4. NP-hard

NP 是指非确定性多项式（Non-deterministic Polynomial，NP）。所谓的非确定性是指，可用一定数量的运算去解决多项式时间内可解决的问题。NP 问题通俗来说是其解的正确性能够被"很容易检查"的问题，这里"很容易检查"指的是存在一个多项式检查算法。若 NP 中所有问题到某一个问题是图灵可归约的，则该问题为 NP-hard 问题。

### 5. K-SVD algorithm

K-SVD 算法是一种经典的字典训练算法，依据误差最小原则，对误差项进行 SVD 分解，选择使误差最小的分解项作为更新的字典原子和对应的原子系数，经过不断的迭代从而得到优化的解。

## Comprehension

### Blank Filling

1. Machine learning is the scientific study of _____ and _____.
2. Machine learning algorithms build a mathematical model of sample data, known as _____.

3. Machine learning algorithms are used in the applications of _____, _____, and _____.

4. Supervised learning has two types: _____ and _____.

5. Training samples for sparse dictionary learning are represented as _____.

6. _____ is one of the predictive modeling approaches used in statistics, data mining and machine learning.

7. _____ is used to discover relationships between variables in large databases.

**Content Questions**

1. What is the main role of data mining?

2. When is a classification algorithm used?

3. What is the regression algorithm named after? What does that mean?

4. What mathematical model does unsupervised learning algorithm establish?

5. In what form does reinforcement learning algorithm give feedback in a dynamic environment?

    What is it used for?

6. Sparse dictionary learning has been applied in image denoising. What is the key idea?

7. What does anomaly detection mean in data mining?

# Answers

**Blank Filling**

1. algorithms; statistical models

2. training data

3. email filtering; detection of network intruders; computer vision

4. classification algorithms; regression algorithms

5. linear combination of basis functions

6. Decision tree

7. Association rule learning

**Content Questions**

1. Data mining mainly conducts exploratory data analysis through unsupervised learning.

2. Classification algorithms are used when the outputs are restricted to a limited set of values.

3. Regression algorithms are named for their continuous outputs, meaning they may have any value within a range.

4. In unsupervised learning, the algorithm builds a mathematical model of a set of data which contains only inputs and no desired outputs.

5. Reinforcement learning algorithms are given feedback in the form of positive or negative reinforcement in a dynamic environment, and are used in autonomous vehicles or in learning to play a game against a human opponent.

6. Sparse dictionary learning has also been applied in image de-noising. The key idea is that a clean image patch can be sparsely represented by an image dictionary, but the noise cannot.

7. In data mining, anomaly detection is the identification of rare items, events or observations which raise suspicions by differing significantly from the majority of the data.

# 参考译文 A

机器学习这个名字是阿瑟·塞缪尔在 1959 年提出的。机器学习是对算法和统计模型的科学研究，计算机系统使用这些算法和统计模型来逐步提高它们在特定任务中的性能。机器学习算法建立样本数据的数学模型，称为"训练数据"，以便在不显式编程执行任务的情况下进行预测或决策。机器学习算法用于电子邮件过滤、网络入侵者检测和计算机视觉等应用中，在这些应用中，开发特定指令的算法来执行任务是不可行的。机器学习与计算统计学密切相关，计算统计学的重点是利用计算机进行预测。数学优化的研究为机器学习领域提供了方法、理论和应用领域。数据挖掘是机器学习中的一个研究领域，主要通过非监督学习进行探索性的数据分析。在其跨业务问题的应用中，机器学习也被称为预测分析。

**机器学习任务**

支持向量机是一种有监督的学习模型，它将数据分成由线性边界分隔的区域。在这里，线性边界把黑圈和白圈分开。

机器学习任务被分为几个大类。在监督学习中，该算法建立了一组数据的数学模型，该数据集包含输入和期望输出。例如，如果任务是确定一个图像是否包含某个对象，监督学习算法的训练数据将包括有和没有该对象（输入）的图像，每个图像将有一个标签（输出），指示它是否包含该对象。在特殊情况下，输入可能只有部分可用，或者仅限于特殊反馈。半监督学习算法从不完整的训练数据中开发数学模型，其中一部分样本输入缺少所需的输出。

分类算法和回归算法是监督学习的两种类型。当输出被限制在一组有限的值时，使用分类算法。对于过滤电子邮件的分类算法，输入将是传入的电子邮件，输出将是要在其中归档电子邮件的文件夹的名称。对于识别垃圾邮件的算法，输出将是"垃圾邮件"或"非垃圾邮件"的预测，由布尔值 1 和 0 表示。回归算法是根据它们的连续输出来命名的，这意味着它们可能在一定范围内具有某些值。连续值的例子是一个物体的温度、长度或价格。

在无监督学习中，该算法建立了一组只包含输入而不包含期望输出的数据的数学模型。无监督学习算法用于在数据中寻找结构，如数据点的分组或聚类。无监督学习可以发现数据中的模式，并可以将输入分组到类别中，就像在特征学习中一样。降维是减少一组数据中"特征"或输入的数量的过程。

主动学习算法对筛选出来的输入数据进行了人工标注，并对标注后的输入数据进行优化，从而获得所需的输出数据。当交互使用时，这些可以呈现给人类用户进行标记。强化学习算法在动态环境中以正强化或负强化的形式给出反馈，用于自动驾驶汽车或学习与人类棋手对弈。机器学习的其他专门算法包括主题建模，在主题建模中，给计算机程序提供一组自然语言文档，并查找包含类似主题的其他文档。在密度估计问题中，机器学习算法可以用来寻找不可观测的概率密度函数。元学习算法根据以往的经验学习自己的归纳偏差。在开发机器人技术中，机器人学习算法产生自己的学习经验序列，也称为课程，通过自主探索和与人类的社会互动，逐步获得新的技能。这些机器人使用指导机制，如主动学习、成熟、运动协同和模仿。

**处理流程和技术**

各种处理流程、技术和方法可应用于一种或多种机器学习算法，以提高其性能。

- 特征学习

特征学习算法，也被称为表示学习算法，通常会保留输入的信息，在进行分类或预测之前按照需要对其进行转换。这种技术允许对来自未知数据生成分布的输入进行重构，并避免在该分布下不合理的配置。这取代了手工特征工程，而且允许机器学习特征并使用它们执行特定的任务。

- 稀疏字典学习

稀疏字典学习是一种特征学习方法，训练样本表示为基函数的线性组合，假设为稀疏矩阵。该方法是一种 NP 难题，难以近似求解。稀疏字典学习的一种常用启发式方法是 K-SVD 算法。它已被应用于多种场合。在分类中，问题是确定以前未见过的训练示例属于哪类。对于已经构建了各类的字典，一个新的训练示例与相应字典最稀疏表示的类相关联。稀疏字典学习在图像去噪中也得到了应用。关键思想是干净的图像补丁可以用图像字典稀疏表示，但是噪声不能。

- 异常检测

在数据挖掘中，异常检测，也称为离群值检测，是指对罕见的项目、事件或观测结果的识别，与大多数数据存在显著差异，从而引起怀疑。通常情况下，这些异常项代表了一个问题，例如银行欺诈、结构缺陷、医疗问题或文本中的错误。异常被称为离群值、新奇性、噪声、偏差和异常。

- 决策树

决策树学习使用决策树作为预测模型，从对项目的观察（在分支中表示）到对项目的目标值的结论（在叶中表示）。它是统计学、数据挖掘和机器学习中常用的预测建模方法之一。目标变量可以取一组离散值的树模型称为分类树；在这些树结构中，叶子表示类标签，而分支表示导致这些类标签的特征的连接。目标变量可以取连续值（通常是实数）的决策树称为回归树。在决策分析中，决策树可以直观、明确地表示决策和决策。在数据挖掘中，决策树描述的是数据，而生成的分类树可以作为决策的输入。

- 关联规则

关联规则学习是一种基于规则的机器学习方法，用于发现大型数据库中变量之间的关系。它的目的是使用一些"趣味性"度量来识别数据库中发现的强规则。这种基于规则的方法在分析更多数据时生成新的规则。假设数据集足够大，最终目标是帮助机器模拟人脑对未分类数据的特征提取和抽象关联能力。

# Text B

## *Cyber Security* and Machine Learning

When it comes to cyber security, no nation can afford to slack off. If a nation's *defense* systems cannot anticipate how an attacker will try to fool them, then an especially clever attack could expose *military* secrets or use disguised malware to cause major networks to crash.

A nation's defense systems must keep up with the *constant* threat of attack, but this is a difficult and never-ending process. It seems that the defense is always playing catch-up.

Ben Rubinstein, a professor at the University of Melbourne in Australia, asks: "Wouldn't it be good if we knew what the malware writers are going to do next, and to know what type of malware is likely to get through the filters?"

In other words, what if defense systems could learn to anticipate how attackers will try to fool them?

**New Words and Expressions**
**cyber security**
  网络安全
**defense**/dɪˈfens/ n.
  防卫，防护；防御措施
**military**/ˈmɪlɪt(ə)rɪ/ adj.
  军事的；军人的
**constant**/ˈkɒnst(ə)nt/ adj
  不变的；恒定的
**adversarial**/ˌædvəˈseərɪəl/ adj.
  对抗的；对手的
**unaided**/ʌnˈeɪdɪd/ adj.
  未受协助的；无助的

Figure 3-6    Cyber Security

## *Adversarial* Machine Learning

In order to address this question, Rubinstein studies how to prepare machine-learning systems to catch adversarial attacks. In the game of national cyber security, these adversaries are often individual hackers or governments who want to trick machine-learning systems for profit or political gain.

Nations have become increasingly dependent on machine-learning systems to protect against such adversaries. *Unaided* by humans, machine-learning systems in anti-malware and *facial recognition* software have the ability to learn and improve their function as they *encounter* new data. As they learn, they become better at catching adversarial attacks.

Machine-learning systems are generally good at catching adversaries, but they are not completely immune to threats, and adversaries are constantly looking for new ways to fool them. Rubinstein says, "Machine learning works well if you give it data like it's seen before, but if you give it data that it's never seen before, there's no guarantee that it's going to work."

With adversarial machine learning, security agencies address this weakness by presenting the system with different types of *malicious* data to test the system's filters. The system then digests this new information and learns how to identify and *capture* malware from clever attackers.

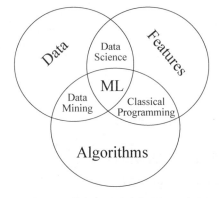

Figure 3-7　Machine Learning in Network Security

### *Security Evaluation* of Machine-Learning Systems

Rubinstein's project is called "Security Evaluation of Machine-Learning Systems", and his *ultimate goal* is to develop a software tool that companies and government agencies can use to test their defenses. Any company or agency that uses machine-learning systems could run his software against their system. Rubinstein's tool would attack and try to fool the system in order to expose the system's *vulnerabilities*. In doing so, his tool anticipates how an attacker could *slip* by the system's defenses.

The software would evaluate existing machine-learning systems and find weak spots that adversaries might try to exploit — similar to how one might defend a castle.

"We're not giving you a new castle," Rubinstein says, "we're

***New Words and Expressions***

**facial recognition**

面部识别

**encounter**/ɪnˈkaʊntə/ v.

遭遇；遇到

**malicious**/məˈlɪʃəs/ adj.

恶意的；恶毒的

**capture**/ˈkæptʃə/ v.

俘获；夺得

**security evaluation**

安全评估

**ultimate goal**

最终目标

**vulnerability**/ˌvʌlnərəˈbɪlətɪ/ n.

易损性；弱点

**slip**/slɪp/ v.

滑动；滑倒

just going to walk around the *perimeter* and look for holes in the walls and weak parts of the castle, or see where the moat is too shallow."

By analyzing many different machine-learning systems, his software program will pick up on trends and be able to advise security agencies to either use a different system or *bolster* the security of their existing system. In this sense, his program acts as a consultant for every machine-learning system.

Consider a program that does facial recognition. This program would use machine learning to identify faces and catch adversaries that pretend to look like someone else.

Rubinstein explains: "Our software aims to automate this security evaluation so that it takes an image of a person and a program that does facial recognition, and it will tell you how to change its appearance so that it will *evade* detection or change the outcome of machine learning in some way."

This is called a mimicry attack — when an *adversary* makes one instance (one face) look like another, and thereby fools a system.

To make this example easier to *visualize*, Rubinstein's group built a program that *demonstrates* how to change a face's appearance to fool a machine-learning system into thinking that it is another face.

When Rubinstein's software fools a system with a mimicry attack, *security personnel* can then take that information and retrain their program to establish more effective security when the stakes are higher.

While Rubinstein's software will help to secure machine-learning systems against adversarial attacks, he has no illusions about the natural advantages that attackers enjoy. It will always be easier to attack a castle than to defend it, and the same holds true for a machine-learning system. This is called the '*asymmetry* of cyber warfare'.

"The attacker can come in from any angle. It only needs to succeed at one point, but the defender needs to succeed at all points," says Rubinstein.

In general, Rubinstein worries that the tools available to test machine-learning systems are theoretical in nature, and put too much responsibility on the *security personnel* to understand the

---

### New Words and Expressions

**perimeter**/pəˈrɪmɪtə/ n.
周长；周界

**bolster**/ˈbəʊlstə/ v.
支持；支撑

**evade**/ɪˈveɪd/ v.
逃避；规避

**adversary**/ˈædvəs(ə)rɪ/ n.
对手；敌手

**visualize**/ˈvɪʒuəlaɪz/ v.
形象，形象化

**demonstrate**/ˈdemənstreɪt/ v.
证明；展示

**asymmetry**/eɪˈsɪmɪtrɪ/ n.
不对称

**security personnel**
安保人员

complex math involved. A researcher might *redo* the mathematical analysis for every new learning system, but security personnel are unlikely to have the time or resources to keep up.

Rubinstein aims to "bring what's out there in theory and make it more applied and more practical and easy for anyone who's using machine learning in a system to *evaluate* the security of their system."

With his software, Rubinstein intends to help level the playing field between attackers and defenders. By giving security agencies better tools to test and adapt their machine-learning systems, he hopes to improve the ability of security personnel to anticipate and *guard against* cyber attacks.

## 参考译文 B

### 网络安全和机器学习

在网络安全方面，任何国家都不能松懈。如果一个国家的国防系统无法预测攻击者将如何试图欺骗它们，那么一次特别熟练的攻击可能会暴露军事机密，或者使用伪装的恶意软件导致主干网络崩溃。

一个国家的国防系统必须跟上受到不断攻击的威胁，但这是一个艰难和永无止境的过程。防守似乎总是在追赶攻击。

澳大利亚墨尔本大学的本·鲁宾斯坦（Ben Rubinstein）教授问道："如果我们知道恶意软件作者接下来会做什么，知道什么样的恶意软件可能会通过过滤器，那该有多好啊！"

换句话说，如果防御系统能够学会预测攻击者将如何试图欺骗他们，那会怎样？

### 敌对的机器学习

为了解决这个问题，鲁宾斯坦研究了如何使用机器学习系统来发现对抗性攻击。在国家网络安全的博弈中，这些对手通常是个人黑客或政府，他们想通过欺骗机器学习系统来获取利润或政治利益。

各国越来越依赖机器学习系统来抵御这类对手。反恶意软件和人脸识别软件中的机器学习系统在没有人类帮助的情况下，能够在遇到新数据时学习和改进自己的功能。随着它们的学习，它们变得更善于发现对手的攻击。

机器学习系统通常善于捕获对手，但它们并不是完全不受威胁的，而且对手不断地寻找新的方法来愚弄它们。鲁宾斯坦说："如果你给它以前见过的数据,机器学习工作得很好,但是如果你给它以前从未见过的数据，就不能保证它会工作。"

在对抗性机器学习中，安全机构通过向系统提供不同类型的恶意数据来测试系统的过滤器以解决这一弱点。然后系统消化这些新信息，学习如何从熟练的攻击者那里识别和捕获恶意软件。

### 机器学习系统的安全性评估

鲁宾斯坦的项目叫做"机器学习系统的安全评估"，他的最终目标是开发一种软件工具，公司和政府机构可以用它来测试他们的防御系统。任何使用机器学习系统的公司或机构都可以在他们的系统上运行其软件。鲁宾斯坦的工具会攻击并试图愚弄系统，以便暴露系统的漏洞。在此过程中，他的工具可以预测攻击者如何从系统的防御中逃脱。

该软件将评估现有的机器学习系统，找出敌人可能试图利用的弱点——类似于一个人如何保卫一座城堡。

鲁宾斯坦说："我们不会给你一座新城堡，我们只是在周边走走，看看墙上有没有洞，城堡的薄弱部分有没有洞，或者看看护城河有没有太浅的地方。"

通过分析许多不同的机器学习系统，他的软件程序将了解趋势，并能够建议安全机构使用不同的系统或加强现有系统的安全性。从这个意义上说，他的程序是每一个机器学习系统的顾问。

考虑做一个面部识别的程序。该程序将使用机器学习识别人脸，并捕捉伪装成其他人的对手。

鲁宾斯坦解释道："我们的软件旨在自动化安全评估，需要一个人的形象和一个程序，面部识别会告诉你如何改变它的外观，这样它会逃避检测或以某种方式改变机器学习的结果。"

这被称为模拟攻击——当对手使一个实例（一张脸）看起来像另一个实例，从而欺骗系统。

为了让这个例子更容易形象化，鲁宾斯坦的团队开发了一个程序，演示如何改变一张脸的外观，从而欺骗机器学习系统，让它以为那是另一张脸。

当鲁宾斯坦的软件通过模拟攻击欺骗系统时，安全人员可以利用这些信息，并在风险较高时重新调整他们的程序，以建立更有效的安全系统。

尽管鲁宾斯坦的软件有助于保护机器学习系统免受对敌攻击，但他也知道攻击者具有天然的优势。攻击城堡总是比保卫城堡容易，机器学习系统也是如此。这就是所谓的"网络战的不对称"。

"攻击者可以从任何角度进入，它只需要在一点上成功，但是防守者需要在所有点上都成功。"鲁宾斯坦说。

总的说来，鲁宾斯坦担心测试机器学习系统的工具本质上是理论性的，并把太多的责任推给了安全人员去理解复杂的数学问题。研究人员可能会为每一个新的学习系统重新进行数学分析，但安全人员不太可能有充足的时间或资源。

鲁宾斯坦的目标是"将现有的理论应用到系统中，让那些在系统中使用机器学习的人更容易、更实用地评估系统的安全性。"

鲁宾斯坦通过他的软件来平衡攻击者和防御者之间的竞争环境。通过给安全机构提供更好的工具来测试和调整他们的机器学习系统，他希望提高安全人员预测和防范网络攻击的能力。

# Chapter 4

# Deep Learning

## Text A

Deep learning (also known as deep structured learning or *hierarchical* learning) is part of a broader family of machine learning methods based on learning data representations, as opposed to *task-specific* algorithms. Learning can be supervised, semi-supervised or unsupervised.

Deep learning architectures such as deep neural networks, deep belief networks and recurrent neural networks have been applied to fields including computer vision, speech recognition, natural language processing, audio recognition, social network filtering, machine translation, bioinformatics, drug design, medical image analysis, *material inspection* and board game programs, where they have produced results comparable to and in some cases superior to human experts.

Deep learning models are *vaguely* inspired by information processing and communication patterns in biological nervous systems yet have various differences from the structural and *functional properties* of biological brains (especially human brains), which make them *incompatible with* neuroscience evidences.

### Overview

Most modern deep learning models are based on an artificial neural network, although they can also include *propositional formulas* or *latent variable*s organized layer-wise in deep generative models such as the nodes in deep belief networks[1] and deep Boltzmann machines.

**New Words and Expressions**

**hierarchical**/haɪəˈrɑːkɪk(ə)l/ adj.
分层的；等级体系的
**task-specific**
任务特异
**material inspection**
材料检验
**vaguely**/veɪgli/ adj.
模糊的；含糊的
**functional property**
功能特性
**incompatible with**
与…不适宜
**propositional**/prɒpəˈzɪʃənl/ adj.
命题的；建议的
**formula**/ˈfɔːmjʊlə/ n.
公式，准则
**latent variable**
潜在变量

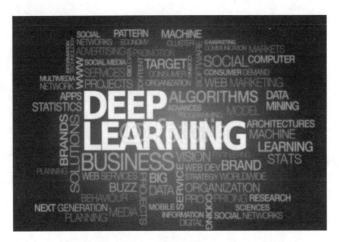

Figure 4-1　Deep Learning

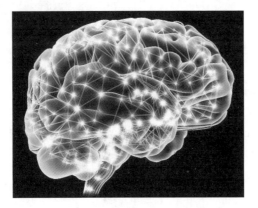

Figure 4-2　Human Brain Neural Network

In deep learning, each level learns to transform its input data into a slightly more abstract and composite representation. In an *image recognition* application, the raw input may be a matrix of *pixels*; the first representational layer may abstract the pixels and encode edges; the second layer may compose and encode arrangements of edges; the third layer may encode a nose and eyes; and the fourth layer may recognize that the image contains a face. Importantly, a deep learning process can learn which features to optimally place in which level on its own. (Of course, this does not completely *obviate* the need for hand-tuning; for example, varying numbers of layers and layer sizes can provide different degrees of abstraction.)

The "deep" in "deep learning" refers to the number of layers through which the data is transformed. More *precisely,* deep learning systems have a substantial credit assignment path (CAP)[2] depth. The CAP is the chain of transformations from input to output. CAPs describe potentially causal connections between input and

*New Words and Expressions*

**image recognition**
图像识别

**pixel**/ˈpɪks(ə)l / n.
像素

**obviate**/ˈɒbvɪeɪt/ v.
排除；避免

**precisely**/prɪˈsaɪslɪ/ adv.
精确地

output. For a *feedforward* neural network[3], the depth of the CAPs is that of the network and is the number of hidden layers plus one (as the output layer is also parameterized). For recurrent neural networks, in which a signal may *propagate* through a layer more than once, the CAP depth is potentially unlimited. No universally agreed upon *threshold* of depth divides shallow learning from deep learning, but most researchers agree that deep learning involves CAP depth > 2. CAP of depth 2 has been shown to be a universal approximator in the sense that it can emulate any function. Beyond that more layers do not add to the function *approximator* ability of the network. Deep models (CAP > 2) are able to extract better features than shallow models and hence.

Deep learning architectures are often constructed with a greedy *layer-by-layer* method. Deep learning helps to *disentangle* these abstractions and pick out which features improve performance.

For supervised learning tasks, deep learning methods obviate feature engineering[4], by translating the data into compact intermediate representations akin to *principal components*, and derive layered structures that remove *redundancy* in representation.

Deep learning algorithms can be applied to unsupervised learning tasks. This is an important benefit because unlabeled data are more abundant than labeled data. Examples of deep structures that can be trained in an unsupervised manner are neural history compressors and deep belief networks.

## Relation to Human Cognitive and Brain Development

Deep learning is closely related to a class of theories of brain development (specifically, *neocortical* development) proposed by *cognitive* neuroscientists in the early 1990s. These developmental theories were *instantiated* in computational models, making them *predecessors* of deep learning systems. These developmental models share the property that various proposed learning dynamics in the brain (e.g., a wave of nerve growth factor[5]) support the self-organization somewhat *analogous* to the neural networks utilized in deep learning models. Like the *neocortex*, neural networks employ a hierarchy of layered filters in which each layer considers information from a prior layer (or the operating environment), and then passes its output (and possibly the original input), to other layers. This process yields a self-organizing stack of transducers, *well-tuned* to their operating environment. A 1995 description stated, "...the

***New Words and Expressions***

**feedforward**/ˈfiːdfɔːwəd/ n.
正反馈

**propagate**/ˈprɒpəɡeɪt/ v.
传播；繁殖

**threshold**/ˈθreʃəʊld/ n.
入口；门槛

**approximator**/əˈprɒksɪmeɪtə/ n.
接近者；合拢器

**layer-by-layer**
叠层

**disentangle**/dɪsɪnˈtæŋɡ(ə)l/ v.
解开；松开

**principal component**
主成分

**redundancy**/rɪˈdʌnd(ə)nsɪ/ n.
冗余

**neocortical**/ˌniːəʊˈkɔːtikəl/ adj.
新（大脑）皮层的

**cognitive**/ˈkɒɡnɪtɪv/ adj.
认知的，认识的

**instantiate**/ɪnˈstænʃɪeɪt/ v.
例示，举例说明

**predecessor**/ˈpriːdɪsesə/ n.
前任，前辈

**analogous**/əˈnæləɡəs/ adj.
类似的

**neocortex**/ˌniːəʊˈkɔːteks/ n.
新（大脑）皮质

**well-tuned**
调整好的

*infant*'s brain seems to organize itself under the influence of waves of so-called *trophic*-factors ... different regions of the brain become connected sequentially, with one layer of *tissue* maturing before another and so on until the whole brain is mature."

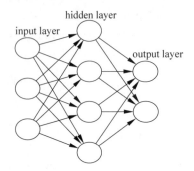

Figure 4-3　Neural Network Structure

A variety of approaches have been used to investigate the *plausibility* of deep learning models from a neurobiological perspective. On the one hand, several *variants* of the *backpropagation* algorithm have been proposed in order to increase its processing realism. Other researchers have argued that unsupervised forms of deep learning, such as those based on hierarchical generative models and deep belief networks, may be closer to biological reality. In this respect, generative neural network models have been related to neurobiological evidence about sampling-based processing in the *cerebral cortex*.

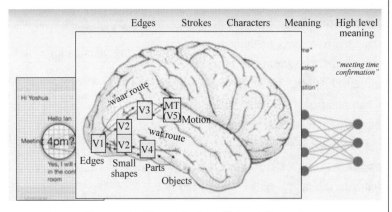

Figure 4-4　Human Brain and Neural Network

Although a *systematic comparison* between the human brain organization and the neuronal encoding in deep networks has not yet been established, several analogies have been reported. For example, the computations performed by deep learning units could

***New Words and Expressions***

**infant**/ˈɪnf(ə)nt/ adj. n.
　婴儿的；幼稚的；初期的；婴儿

**trophic**/ˈtrəʊfɪk/ adj.
　营养的；有关营养的

**tissue**/ˈtɪʃuː/ n.
　组织；纸巾

**plausibility**/ˌplɔːzəˈbɪlətɪ/ n.
　善辩；似乎有理

**variant**/ˈveərɪənt/ n.
　变体；转化

**backpropagation** n.
　反向传播

**cerebral**/ˈserɪbr(ə)l/ adj.
　大脑的，脑的

**cortex**/ˈkɔːteks/ n.
　皮质；树皮

**systematic comparison**
　系统比较

be similar to those of actual neurons and neural populations. Similarly, the representations developed by deep learning models are similar to those measured in the primate visual system both at the *single-unit* and at the population levels.

**New Words and Expressions**

**single-unit**

单机制；单一机组

## Terms

### 1. Deep Belief Networks (DBN)

深度信念网络由 Geoffrey Hinton 在 2006 年提出。它是一种生成模型，通过训练其神经元间的权重，我们可以让整个神经网络按照最大概率来生成训练数据。我们不仅可以使用 DBN 识别特征、分类数据，还可以用它来生成数据。

### 2. Credit Assignment Path (CAP)

贡献度分配问题，这是深度学习中的一个重要问题，即一个系统中不同的组件（Components）或其参数对最终系统输出结果的贡献或影响，它关系到对参数的学习。目前在深度学习中，神经网络能较好地解决这个问题。

### 3. Feedforward Neural Network

前馈神经网络是一种最简单的神经网络，各神经元分层排列。每个神经元只与前一层的神经元相连。接收前一层的输出，并输出给下一层，各层间没有反馈。前馈神经网络是目前应用最广泛、发展最迅速的人工神经网络之一。研究从 20 世纪 60 年代开始，目前理论研究和实际应用达到了很高的水平。

### 4. Feature Engineering

特征工程指的是把原始数据转变为模型的训练数据的过程，它的目的就是获取更好的训练数据特征，使得机器学习模型逼近这个上限。特征工程能使得模型的性能得到提升，有时甚至在简单的模型上也能取得不错的效果。

### 5. Nerve Growth Factor (NGF)

神经生长因子能促进中枢和外周神经元的生长、发育、分化、成熟，维持神经系统的正常功能，加快神经系统损伤后的修复。NGF 广泛分布于机体各组织器官中（包括脑），在靶组织中的浓度与交感神经和感觉神经在靶区分支的密度和 mRNA 的含量有关。

## Comprehension

### Blank Filling

1. Deep learning is part of a broader family of machine learning methods based on _____ _____, as opposed to task-specific algorithms.

2. Deep learning models are vaguely inspired by information processing and communication patterns in _____.

3. Most modern deep learning models are based on an _____.

4. Deep learning architectures are often constructed with a _____.

5. Deep learning is closely related to a class of theories of _____ proposed by cognitive neuroscientists.

**Content Questions**

1. What are the application areas of deep learning?

2. What should each level learn to do in deep learning?

3. What does "deep" mean in "deep learning"?

4. How does the deep learning approach avoid feature engineering for supervised learning tasks?

5. What is the common feature of the brain development model?

# Answers

**Blank Filling**

1. learning data representations

2. biological nervous systems

3. artificial neural network

4. greedy layer-by-layer method

5. brain development

**Content Questions**

1. It include computer vision, speech recognition, natural language processing, audio recognition, social network filtering, machine translation, bioinformatics, drug design, medical image analysis, material inspection and board game programs.

2. In deep learning, each level learns to transform its input data into a slightly more abstract and composite representation.

3. The "deep" in "deep learning" refers to the number of layers through which the data is transformed.

4. For supervised learning tasks, deep learning methods obviate feature engineering, by translating the data into compact intermediate representations akin to principal components, and derive layered structures that remove redundancy in representation.

5. These developmental models share the property that various proposed learning dynamics in the brain support the self-organization somewhat analogous to the neural networks utilized in deep learning models.

# 参考译文 A

深度学习（也称为深度结构化学习或分层学习）是基于学习数据表示的更广泛的机器学习方法的一部分，而不是特定于任务的算法。深度学习可以是有监督的、半监督的或无监督的。

深度学习架构（如深度神经网络、深度信念网络和循环神经网络）的应用领域包括计算机视觉、语音识别、自然语言处理、音频识别、社交网络过滤、机器翻译、生物信息学、药物设计、医学图像分析、材料检验和棋盘游戏项目，它们产生了相当大的成效，在某些情况下优于人类专家。

深度学习模型的灵感来源于生物神经系统的信息处理和通信模式，但与生物大脑（尤其是人脑）的结构和功能特性存在着诸多差异，与神经科学已经证明的生物事实不一致。

### 概述

大多数现代的深度学习模型都是基于人工神经网络，尽管它们也可以包含命题公式或潜在变量，这些潜在变量在深层生成模型中按层组织，比如在深度信念网络中的节点和深层玻尔兹曼机中。

在深度学习中，每个层次都学习将其输入数据转换为稍微抽象和复合的表示形式。在图像识别应用程序中，原始输入可以是像素矩阵；第一表示层可以对像素进行提取并对边缘进行编码；第二层可对边缘的排列进行组合和编码；第三层可能编码鼻子和眼睛；第四层可能识别出图像中包含一个人脸。重要的是，深度学习过程可以自己学习哪些特性最适合放在哪个级别。（当然，这并不完全排除手工调优的需要。例如，不同数量的层和层大小可以提供不同的抽象程度。）

"深度学习"中的"深度"指的是数据转换的层数。更准确地说，深度学习系统具有深度的贡献度分配路径（CAP）。贡献度分配是从输入到输出的转换链，它描述了输入和输出之间潜在的因果关系。对于前馈神经网络，贡献度分配的深度是网络的深度，是隐藏层的数量加 1（因为输出层也是参数化的）。对于递归神经网络，信号可能不止一次通过一层传播，其贡献度分配深度可能是无限的。目前还没有公认的深度阈值将浅学习与深度学习区分开来，但大多数研究者认为深度学习涉及到 CAP > 2。深度贡献度分配 2 已经被证明是一个通用的近似器，在这个意义上，它可以模拟任何函数。除此之外，更多的层次并不会增加网络的函数逼近能力。深度模型（CAP > 2）能够比浅层模型提取更好的特征，因此，额外的层有助于学习特征。

深度学习体系结构通常使用逐层的贪心算法构建。深度学习有助于理清这些抽象概念，并找出哪些特性可以提高性能。

对于监督学习任务，深度学习方法通过将数据转换为类似于主成分的紧凑的中间表示，并推导出消除表示冗余的分层结构，从而避免特征工程。

深度学习算法可以应用于无监督学习任务。这是一个重要的优点，因为未标记的数据比标记的数据更丰富。可以在无监督方式下进行训练的深层结构的例子有神经历史压缩器和深层信念网络。

### 与人类认知和大脑发育的关系

深度学习与认知神经科学家在 20 世纪 90 年代初提出的大脑发展（特别是新皮层发展）理论密切相关。这些发展理论在计算模型中得到了例证，成为深度学习系统的前身。这些发展模型与促进大脑学习动力的因素具有相同的特征（如神经生长因子的脑波），在某种程度上类似于深度学习模型使用的神经网络。与新大脑皮层一样，神经网络采用分层过滤的层次结构，每一层都考虑来自前一层（或操作环境）的信息，然后将其输出（可能还有原始输入）传递给其他层。这个过程产生一个自组织的传感器堆栈，很好地适应了它们的操作环境。1995 年的一份描述说："……婴儿的大脑似乎在所谓的营养因子波的影响下自行组织起来……大脑的不同区域依次连接，一层组织先于另一层组织成熟，以此类推，直到整个大脑成熟。"

从神经生物学的角度研究深度学习模型的合理性，已经使用了多种方法。为了提高反向传播算法的处理真实感，提出了几种不同的反向传播算法。其他研究人员认为，非监督形式的深度学习，例如基于层次生成模型和深度信念网络的深度学习，可能更接近生物学现实。在这方面，生成神经网络模型已经与大脑皮层基于采样处理的神经生物学证据相关。

虽然目前还没有系统地比较人脑组织和深层网络中的神经编码，但是已经有一些类似的报道。例如，由深度学习单元进行的计算可以类似于实际神经元和神经种群的计算。同样地，由深度学习模型发展出来的表象，无论是在单个单位还是在种群水平上，都与灵长类视觉系统测量到的表象相似。

# Text B

Imagine how much more efficient lawyers could be if they had the time to read every legal book ever written and review every case ever brought to court. Imagine doctors with the ability to study every advancement published across the world's medical journals, or consult every medical case, ever. Unfortunately, the human brain cannot store that much information, and it would take decades to achieve these feats.

But a computer, one specifically designed to work like the human mind, could.

| New Words and Expressions |
| :--- |
| **mimic**/ˈmɪmɪk/ v. |
| 模仿，摹拟 |
| **exposure**/ɪkˈspəʊʒə/ n. |
| 暴露；曝光 |

Figure 4-5    Artificial Intelligence

Deep learning neural networks are designed to *mimic* the human brain's neural connections. They are capable of learning through continuous *exposure* to huge amounts of data. This allows them to recognize patterns, comprehend complex concepts, and translate high-level abstractions. These networks consist of many layers, each having a different set of weights. The deeper the network, the stronger it is.

Current applications for these networks include medical diagnosis, robotics and engineering, face recognition, and automotive navigation. However, deep learning is still in development — not surprisingly, it is a huge *undertaking* to get machines to think like humans. In fact, very little is understood about these networks, and months of *manual* tuning are often required for obtaining excellent performance.

**New Words and Expressions**

**undertaking**/ˌʌndəˈteɪkɪŋ/ n.

事业；企业

**manual**/ˈmænjʊ(ə)l/ adj.

手工的；体力的

**convolutional**/ˌkɒnvəˈluːʃ(ə)nəl/ n.

卷积；回旋

**safeguard**/ˈseɪfgɑːd/ n.

保护；保卫

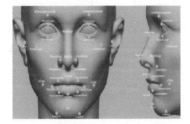

Figure 4-6    Face Recognition

Fuxin Li, assistant professor at the Oregon State University School of Electrical Engineering and Computer Science, and his team are taking on the accuracy of these neural networks under adversarial conditions. Their research focuses on the basic machine learning aspects of deep learning, and how to make general deep learning more robust.

To try to better understand when a deep *convolutional* neural network (CNN) is going to be right or wrong, Li's team had to establish an estimate of confidence in the predictions of the deep learning architecture. Those estimates can be used as *safeguards* when utilizing the networks in real life.

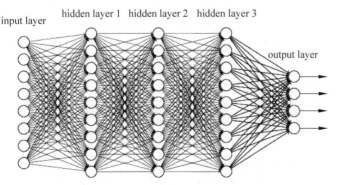

Figure 4-7    Convolutional Neural Network

"Basically," explains Li, "trying to make deep learning increasingly self-aware to be aware of what type of data it has seen, and what type of data it could work on."

The team looked at recent advances in deep learning, which have greatly improved the capability to recognize images automatically. Those networks, *albeit* very resistant to *overfitting*, were discovered to completely fail if some of the pixels in such images were *perturbed* via an *adversarial* optimization algorithm.

To a human observer, the image in question may look fine, but the deep network sees otherwise. According to the researchers, those adversarial examples are dangerous if a deep network is utilized into any *crucial* real application, such as autonomous driving. If the result of the network can be hacked, wrong authentications and other devastating effects would be unavoidable.

In a departure from previous perspectives that focused on improving the *classifiers* to correctly organize the adversarial examples, the team focused on detecting those adversarial examples by analyzing whether they come from the same distribution as the normal examples. The accuracy for detecting adversarial examples exceeded 96%. Notably, 90% of the adversarials can be detected with a false positive rate of less than 10%.

The benefits of this research are numerous. It is vital for a neural network to be able to identify whether an example comes from a normal or an adversarial distribution. Such knowledge, if available, will help significantly to control behaviors of robots employing deep learning. A reliable procedure can prevent robots from behaving in an undesirable manner because of the false perceptions it made about the environment.

Li gives one example: "In robotics there's this big issue about robots not doing something based on erroneous perception. It's important for a robot to know that it's not making a confident perception. For example, if [the robot] is saying there's an object over there, but it's actually a wall, he'll go to *fetch* that object, and then he hits a wall."

Hopefully, Li says, that won't happen. However, current software and machine learning have been mostly based *solely* on prediction confidence within the original machine learning framework. Basically, the testing and training data are assumed to be pulled from

**New Words and Expressions**

**albeit** /ɔːlˈbiːɪt/ conj.
虽然；即使

**overfitting**
过适（overfit 现在分词）

**perturb** /pəˈtɜːb/ v.
扰乱；使…混乱

**adversarial** /ˌædvəˈseərɪəl/ adj.
对抗的；对手的

**crucial** /ˈkruːʃ(ə)l/ adj.
重要的；决定性的

**classifier** /ˈklæsɪfaɪə/ n.
分类器

**fetch** /fetʃ/ v.
取来；接来

**solely** /ˈsəʊllɪ/ adv.
单独地，唯一地

the same distribution independently, and that can lead to incorrect assumptions.

　　Better confidence estimates could potentially help avoid incidents such as the Tesla crash *scenario* from May 2016, where an adversarial example (truck with too much light) was in the middle of the highway that cheated the system. A confidence estimate could potentially solve that issue. But first, the computer must be smarter. The computer has to learn to detect objects and differentiate, say, a tree from another vehicle.

**New Words and Expressions**

**scenario**/sɪˈnɑːrɪəʊ/ n.

　方案；情节

**procedure**/prəˈsiːdʒə/ n.

　程序，手续；步骤

**utilize**/ˈjuːtɪˌlaɪz/ v.

　利用

**self-aware**/ˌselfəˈwɛə/ adj.

　有自知之明的

Figure 4-8　Tesla Crash Scenario

　　"To make it really robust, you need to account for unknown objects. Something weird may hit you. A deer may jump out." The network can't be taught every unexpected situation, says Li, "so you need it to discover them without knowledge of what they are. That's something that we do. We try to bridge the gap."

　　Training *procedures* will make deep learning more automatic and lead to fewer failures, as well as confidence estimates when the deep network is *utilized* to predict new data. Most of this training, explains Li, comes from photo distribution using stock images. However, these are flat images much different than what a robot would normally see in day-to-day life. It's difficult to get a 360-degree view just by looking at photos.

　　"There will be a big difference between the thing [the robot] trains on and the thing it really sees. So then, it is important for the robot to understand that it can predict some things confidently, and others it cannot," says Li. "[The robot] needs to understand that it probably predicted wrong, so as not to act too aggressively toward its prediction." This can only be achieved with a more *self-aware* framework, which is what Li is trying to develop with this grant.

Further, these estimates can be used to control the behavior of a robot employing deep learning so that it will not go on to perform *maneuvers* that could be dangerous because of erroneous predictions. Understanding these aspects would also be helpful in designing potentially more robust networks in the future.

Soon, Li and his team will start *generalizing* the approach to other domains, such as temporal models (RNNs, LSTMs) and deep reinforcement learning. In reinforcement learning, the confidence estimates could play an important role in many decision-making *paradigms*.

> **New Words and Expressions**
>
> **maneuver**/məˈnʊvə/ n.
>
> 　机动；演习
>
> **generalize**/ˈdʒɛnrəˌlaɪz/ v.
>
> 　概括；推广
>
> **paradigm**/ˈpærədaɪm/ n.
>
> 　范例；词形变化表

## 参考译文 B

想象一下，如果律师有时间阅读每本写过的法律书籍，审查每宗提交法庭的案件，他们会变得多么高效。再想象一下，医生有能力研究世界各地医学杂志上发表的每项进展，或咨询每个医学案例。不幸的是，人类的大脑不能储存那么多的信息，要实现这些壮举需要几十年的时间。

但是一台专门设计成像人类思维一样工作的计算机却可以。

深度学习神经网络被设计用来模拟人脑的神经连接。他们能够通过持续接触大量数据来学习。这允许他们识别模式，理解复杂的概念，并解释深层抽象数据。这些网络由许多层组成，每一层都有不同的权重集。网络越深，它就越强大。

目前这些网络的应用包括医疗诊断、机器人和工程、人脸识别和汽车导航。然而，深度学习仍处于发展阶段——这并不奇怪，让机器像人类一样思考是一项艰巨的任务。事实上，人们对这些网络知之甚少，通常需要数月的手工调优才能获得出色的性能。

俄勒冈州立大学电气工程与计算机科学学院助理教授李富新（Fuxin Li）和他的团队正在研究这些神经网络在对抗条件下的准确性。他们的研究集中在深度学习的基础机器学习方面，以及如何使一般的深度学习更加健全。

为了更好地理解深度卷积神经网络（CNN）是对是错，李的团队必须对深度学习架构的预测建立一个信心评估。这些估计可以作为在现实生活中利用网络时的保障。

李解释说，"基本上，试着让深度学习越来越具有自我意识——意识到它看到了什么类型的数据，以及它可以处理什么类型的数据。"

研究小组观察了深度学习的最新进展，深度学习极大地提高了自动识别图像的能力。这些网络虽然能够对抗过拟合，但如果通过对抗性优化算法对图像中的某些像素进行扰动，这些网络就会完全失效。

有的图像人肉眼看可能很好，但深层网络却不这么认为。根据研究人员的说法，如果一个深层网络被用于任何关键的实际应用，例如自动驾驶，那么这些对抗性的例子是危险的。如果网络的结果可以被黑客攻击，错误的认证和其他破坏性的影响将不可避免。

与以前通过改进分类器来对抗样本进行异常检测不同，团队关注于通过分析这些对抗样本是否来自与正常样本相同的分布来检测这些对抗样本。对抗样本的检测准确率超过96%。值得注意的是，90%的对抗可以检测到，而未检出的概率低于10%。

这项研究的好处是多方面的。神经网络能够识别一个例子是来自正态分布还是反向分布是至关重要的。应用这些知识将大大有助于控制使用深度学习的机器人的行为。一个可靠的程序可以防止机器人由于对环境的错误感知而做出错误的行为。

李举了一个例子："在机器人技术中，有一个很大的问题是机器人不能基于错误的感知去做某件事。对机器人来说，重要的是要知道它没有做出自信的感知。例如，如果（机器人）说那边有一个物体，但它实际上是一堵墙，它就会去取那个物体，然后撞上墙。

"希望这不会发生，"李说。然而，目前的软件和机器学习大多是基于原始机器学习框架内的预测置信度。基本上，测试和训练数据被认为是从同一个分布中独立提取的，这可能导致错误的假设。

更好的置信度估计可能有助于避免 2016 年 5 月发生的特斯拉（Tesla）撞车事故，当时，一辆对抗式卡车（载重量过大的卡车）行驶在公路中间，骗过了系统。置信度估计有可能解决这个问题。但首先，计算机必须更聪明。计算机必须学会探测物体并将树与其他交通工具区分开来。

"为了使它完善，你需要考虑未知的对象。"可能会有奇怪的事情发生，比如会有随时跳出来的鹿。我们不可能把任何一种意想不到的情况都教授给网络，"李说，"所以需要在它不知道会是什么的情况下，却能够识别出它们。"我们就是这么做的。我们试图弥合这一差距。

训练过程会使深度学习更加自动化，失败次数更少，利用深度网络预测新数据时，还会进行置信度估计。李解释说，大部分训练来自于库存图片。然而，这些平面图像与机器人通常在日常生活中看到的图像有很大不同。仅仅通过看照片很难获得 360° 的视角。

"（机器人）训练的东西和它真正看到的东西将有很大区别。因此，重要的是机器人要明白，它可以自信地预测一些事情，而不能预测其他事情。（机器人）需要明白，它的预测可能是错误的，这样就不会对自己的预测采取过于激进的行动。"这只能通过一个更有自我意识的框架来实现，而这正是李教授试图用这笔赠款来研发的。

此外，这些估计可以用来控制使用深度学习的机器人的行为，这样它就不会继续执行由于错误预测而可能造成危险的行动。了解这些方面也有助于设计未来可能更加健全的网络。

不久，李和他的团队将开始将这种方法推广到其他领域，例如时间模型（RNN、LSTM）和深度强化学习。在强化学习中，置信度估计在许多决策范式中发挥着重要作用。

# Chapter *5*

## Natural Language Processing

# Text A

Natural language processing (NLP) is one of the most important technologies of the information age, and a *crucial* part of artificial intelligence. Applications of NLP are everywhere because people communicate almost everything in language: web search, advertising, emails, customer service, language translation, medical reports, etc. In recent years, Deep Learning approaches have obtained very high performance across many different NLP tasks, using single end-to-end neural models that do not require traditional, *task-specific* feature engineering.

*New Words and Expressions*

**crucial**/ˈkruːʃ(ə)l/ adj.

重要的；决定性的

**task-specific**

特定任务

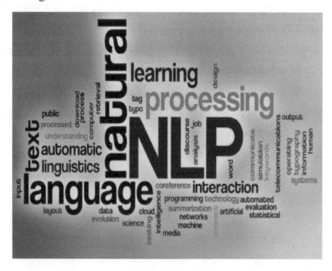

Figure 5-1　Natural Language Processing

### What is Natural Language Processing?

NLP is a way for computers to analyze, understand, and *derive* meaning from human language in a smart and useful way. By utilizing NLP, developers can organize and structure knowledge to perform tasks such as automatic *summarization*, translation, named entity recognition, relationship extraction, *sentiment* analysis, speech recognition, and topic *segmentation*.

Figure 5-2　*Linguistics* in Natural Language Processing

"Apart from common word processor operations that treat text like a mere sequence of symbols, NLP considers the *hierarchical* structure of language: several words make a *phrase*, several phrases make a sentence and, *ultimately*, sentences convey ideas," John Rehling, an NLP expert at Meltwater Group, said in How Natural Language Processing Helps Uncover Social Media Sentiment. "By analyzing language for its meaning, NLP systems have long filled useful roles, such as correcting grammar, *converting* speech to text and automatically translating between languages."

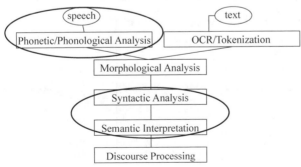

Figure 5-3　Levels Involved in Natural Language Processing

**New Words and Expressions**

**derive**/dɪˈraɪv/ v.

源于；起源

**summarization**/ˌsʌmərɪˈzeɪʃən/ n.

摘要；概要

**sentiment**/ˈsentɪm(ə)nt/ n.

感情，情绪

**segmentation**/ˌseɡmənˈteɪʃən/ n.

分割；割断

**linguistic**/lɪŋˈɡwɪstɪk/ adj.

语言的；语言学的

**hierarchical**/haɪəˈrɑːkɪk(ə)l/ adj.

分层的；等级体系的

**phrase**/freɪz/ n.

短语，习语

**ultimately**/ˈʌltɪmətlɪ/ adv.

最后；根本

**convert**/kənˈvɜːt/ v.

转变，变换

NLP is used to analyze text, allowing machines to understand how human's speak. This human-computer interaction enables real-world applications like automatic text summarization, sentiment analysis, topic *extraction*, named entity recognition, *parts-of-speech* tagging, relationship extraction, *stemming*, and more. NLP is commonly used for *text mining*[1], machine translation, and automated question answering.

NLP is characterized as a hard problem in computer science. Human language is rarely precise, or *plainly* spoken. To understand human language is to understand not only the words, but the concepts and how they're linked together to create meaning. Despite language being one of the easiest things for humans to learn, the *ambiguity* of language is what makes natural language processing a difficult problem for computers to master.

**New Words and Expressions**

**extraction**/ɪkˈstrækʃ(ə)n/ n.
　取出；抽出
**parts-of-speech**
　词类；词性
**stemming**/ˈstemɪŋ/ v.
　词干提取
**text mining**
　文本挖掘
**plainly**/ˈpleɪnli/ adv.
　明白地；坦率地
**ambiguity**/æmbɪˈgjuːɪtɪ/ n.
　含糊；不明确
**syntactic**/sɪnˈtæktɪk/ adj.
　句法的；语法的
**corpus**/ˈkɔːpəs/ n.
　语料库
**irrelevant**/ɪˈrelɪv(ə)nt/ adj.
　不相干的；不切题的

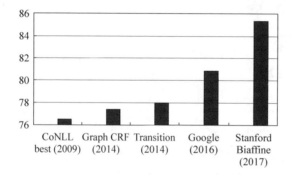

Figure 5-4　*Syntactic* Analysis Performance

## What Can Developers Use NLP Algorithms For?

NLP algorithms are typically based on machine learning algorithms. Instead of hand-coding large sets of rules, NLP can rely on machine learning to automatically learn these rules by analyzing a set of examples (i.e. a large *corpus*, like a book, down to a collection of sentences), and making a statical inference. In general, the more data analyzed, the more accurate the model will be.

- Summarize blocks of text using Summarizer to extract the most important and central ideas while ignoring *irrelevant* information.

- Create a chat bot using Parsey McParseface, a language parsing deep learning model made by Google that uses Point-of-Speech tagging.

- Automatically generate keyword tags from content using AutoTag, which *leverages* LDA[2], a technique that discovers topics contained within a body of text.
- Identify the type of entity extracted, such as it being a person, place, or organization using Named Entity Recognition.
- Use Sentiment Analysis to identify the sentiment of a string of text, from very negative to neutral to very positive.
- Reduce words to their root, or stem, using Porter Stemmer[3], or break up text into tokens using *tokenizer*.

## Example Natural Language Processing Use Cases

Social media analysis is a great example of NLP use. Brands track conversations online to understand what customers are saying, and *glean* insight into user behavior.

"One of the most *compelling* ways NLP offers valuable intelligence is by tracking sentiment — the tone of a written message (tweet, Facebook update, etc.) — and tag that text as positive, negative or neutral," Rehling said.

## Build Your Own Social Media Monitoring Tool

Start by using the algorithm *Retrieve* Tweets With Keyword to capture all mentions of your brand name on Twitter. In our case, we search for mentions of Algorithmia.

Then, pipe the results into the Sentiment Analysis algorithm, which will assign a sentiment rating from 0-4 for each string (Tweet).

Similarly, Facebook uses NLP to track trending topics and popular *hashtags*. "Hashtags and topics are two different ways of grouping and participating in conversations," Chris Struhar, a software engineer on News Feed, said in How Facebook Built Trending Topics with Natural Language Processing. "So don't think Facebook won't recognize a string as a topic without a hashtag in front of it. Rather, it's all about NLP: natural language processing. Ain't nothing natural about a hashtag, so Facebook instead parses strings and figures out which strings are referring to nodes — objects in the network."

It's not just social media that can use NLP to it's benefit. Publishers are hoping to use NLP to improve the quality of their online communities by leveraging technology to "*auto-filter* the *offensive* comments on news sites to save *moderators* from what

**New Words and Expressions**

**leverage**/ˈlev(ə)rɪdʒ/ v.
利用
**tokenizer**/ˈtəʊkənaɪzə/
分词器
**glean**/gliːn/ v.
收集
**compel**/kəmˈpel/ v.
强迫，迫使
**retrieve** /rɪˈtriːv/ v.
检索
**hashtag**/ˈhæʃˌtæg/ n.
主题标签
**auto-filter**
自动筛选
**offensive**/əˈfensɪv/ adj.
攻击的；冒犯的
**moderator**/ˈmɒdəreɪtə/ n.
版主；主持人
**journalism**/ˈdʒɜːn(ə)lɪz(ə)m/ n.
新闻业，新闻工作

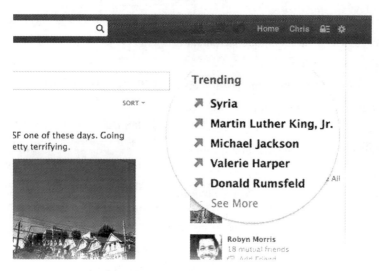

Figure 5-5    Trending Topics on Facebook

**New Words and Expressions**
**malicious**/məˈlɪʃəs/ adj.
恶意的；恶毒的

can be an 'exhausting process'," Francis Tseng said in Prototype winner using "natural language processing" to solve *journalism*'s commenting problem.

Other practical uses of NLP include monitoring for *malicious* digital attacks, such as phishing, or detecting when somebody is lying.

## Terms

### 1. Text Mining

文本挖掘（Text Mining）指的是从文本数据中获取有价值的信息和知识，它是数据挖掘中的一种方法。文本挖掘最重要最基本的应用是实现文本的分类和聚类，前者是有监督的挖掘算法，后者是无监督的挖掘算法。文本挖掘是一个多学科混杂的领域，涵盖了多种技术，包括数据挖掘技术、信息抽取、信息检索、机器学习、自然语言处理、计算语言学、统计数据分析、线性几何、概率理论甚至还有图论。

### 2. LDA

LDA（Latent Dirichlet Allocation）是一种文档主题生成模型，也称为三层贝叶斯概率模型，包含词、主题和文档三层结构。所谓生成模型，就是说，我们认为一篇文章的每个词都是通过"以一定概率选择了某个主题，并从这个主题中以一定概率选择某个词语"这样一个过程得到。文档到主题服从多项式分布，主题到词服从多项式分布。

### 3. Porter Stemmer

英文分词算法（Porter Stemmer）主要可以实现单词原型的还原，一般分为三个步骤：①根据空格/符号/段落分隔，得到单词组；②过滤，排除掉停用词；③提取词干。

## Comprehension

**Blank Filling**

1. Deep Learning approaches have obtained very high performance across many different NLP tasks, using _____ neural models.

2. NLP is a way for computers to _____, _____ , and _____ from human language in a smart and useful way.

3. A common word processor operation in NLP is to treat text like a mere sequence of _____.

4. NLP is used to _____, allowing machines to understand how human's speak.

5. NLP algorithms are typically based on _____ algorithms.

6. "One of the most compelling ways NLP offers valuable intelligence is by _____— the tone of a written message— and tag that text as positive, negative or neutral," Rehling said.

**Content Questions**

1. By utilizing NLP, what tasks can developers organize and structure knowledge to perform?

2. By analyzing language for its meaning, what are some examples of NLP systems that have long filled useful roles?

3. What real-world applications does human-computer interaction enable?

4. What does it take to understand human language?

5. What are the steps required to build your own social media monitoring tool?

6. What are publishers hoping to do with NLP technology?

## Answers

**Blank Filling**

1. single end-to-end

2. analyze; understand; derive meaning

3. symbols

4. analyze text

5. machine learning

6. tracking sentiment

**Content Questions**

1. By utilizing NLP, developers can organize and structure knowledge to perform tasks such as automatic summarization, translation, named entity recognition, relationship extraction, sentiment analysis, speech recognition, and topic segmentation.

2. By analyzing language for its meaning, NLP systems have long filled useful roles, such

as correcting grammar, converting speech to text and automatically translating between languages.

3. This human-computer interaction enables real-world applications like automatic text summarization, sentiment analysis, topic extraction, named entity recognition, parts-of-speech tagging, relationship extraction, stemming, and more.

4. To understand human language is to understand not only the words, but the concepts and how they're linked together to create meaning.

5. Start by using the algorithm Retrieve Tweets With Keyword to capture all mentions of your brand name on Twitter. Then, pipe the results into the Sentiment Analysis algorithm, which will assign a sentiment rating from 0-4 for each string (Tweet).

6. Publishers are hoping to use NLP to improve the quality of their online communities by leveraging technology to auto-filter the offensive comments on news sites to save moderators from what can be an "exhausting process".

## 参考译文 A

自然语言处理是信息时代最重要的技术之一，是人工智能的重要组成部分。自然语言处理的应用无处不在，因为人们几乎用语言交流一切：网络搜索、广告、电子邮件、客户服务、语言翻译、医学报告等。近年来，深度学习方法在许多不同的自然语言处理任务中都获得了非常高的性能，通过使用单一的端到端神经模型，不需要传统的、特定于任务的工程特性。

**什么是自然语言处理？**

自然语言处理是计算机从人类语言中分析、理解和获得意义的一种聪明而有用的方法。利用自然语言处理，开发人员可以组织和结构化知识来执行自动摘要、翻译、命名实体识别、关系提取、情感分析、语音识别和主题分割等任务。

"除了常见的字处理程序操作，把文字视为符号序列，自然语言处理还考虑语言的层次结构：一些单词短语，几个短语造一个句子，最终，句子表达想法。" John Rehling，一位 Meltwater 小组的自然语言处理专家，在关于自然语言处理如何帮助揭示社交媒体表示情绪时说道，"通过分析语言的意义，自然语言处理系统长期以来一直扮演着重要的角色，如纠正语法、将语音转换为文本以及在语言之间自动翻译。"

自然语言处理用于分析文本，使机器能够理解人类的语言。这种人机交互支持实际应用程序，如自动文本摘要、情感分析、主题提取、命名实体识别、词性标注、关系提取、词干提取等。自然语言处理通常用于文本挖掘、机器翻译和自动回答问题。

自然语言处理是计算机科学中的一个难题。人类的语言很少是精确的，或者说清楚的。要理解人类语言，不仅要理解单词，还要理解概念以及它们是如何联系在一起来创造意义的。尽管语言是人类最容易学习的东西之一，但语言的模糊性使自然语言处理成为计算机难以掌握的难题。

**开发人员可以使用自然语言处理算法做什么？**

自然语言处理算法通常基于机器学习算法。自然语言处理不需要手工编码大量的规则，它可以依靠机器学习，通过分析一组例子（例如，一个大的语料库，像一本书，归结为一组句子），进行静态推理，自动学习这些规则。一般来说，分析的数据越多，模型就越准确。

- 使用摘要器总结文本块，以提取最重要和中心的思想，同时忽略无关的信息。
- 使用 Parsey McParseface 创建聊天机器人，Parsey McParseface 是一种由谷歌制作的语言分析深度学习模型，它使用了言语点标记。
- 使用自动标签从内容自动生成关键字标记，自动标签利用文档主题生成模型，这种技术可以发现文本正文中包含的主题。
- 识别提取的实体的类型，例如使用命名实体识别的人、地点或组织。
- 用情绪分析来识别一串文字的情绪，从非常消极到中性再到非常积极。
- 使用英文分词算法将单词减少到词根或词干，或者使用记号赋予器将文本分解为记号。

**自然语言处理应用案例**

社交媒体分析是自然语言处理使用的一个很好的例子。品牌跟踪在线对话，以了解客户在说什么，并收集对用户行为的洞察。

Rehling 说：“通过追踪情绪，自然语言处理可以提供有价值的信息——书面信息（Twitter、Facebook 更新等）的语气——并将其标记为积极的、消极的或中性的。”

**建立自己的社交媒体监控工具**

首先，使用这个算法来检索带有关键字的字符串，以在 Twitter 上捕捉所有提到您品牌的信息。在我们的例子中，我们寻找提到的算法。

然后，将结果导入情绪分析算法，该算法将为每个字符串（Tweet）分配 0～4 的情绪评级。

类似地，Facebook 使用自然语言处理来跟踪热门话题和流行话题标签。“标签和话题是两种不同的分组和参与对话的方式。”信息流的软件工程师克里斯·斯特鲁哈尔（Chris Struhar）在《Facebook 如何用自然语言处理来构建热门话题》一书中说。“所以，不要以为 Facebook 不会把一个字符串识别成一个没有标签的主题。而是关于 NLP：自然语言处理。标签不是自然的东西，所以 Facebook 会解析字符串并找出哪些字符串引用了网络中的节点——对象。”

不仅仅是社交媒体可以利用自然语言处理为自己谋利。出版商希望利用自然语言处理技术，“自动过滤新闻网站上的攻击性评论，使版主免于‘累人的过程’，从而提高其在线社区的质量。”Francis Tseng 说。

自然语言处理的其他实际用途包括监视恶意数字攻击，如钓鱼或检测某人何时说谎。

# Text B

## Using Natural Language Processing to Identify Malicious Domains

　　*Cybercriminals* apparently have a tendency to use the same (or at least similar) *lexical* styles when establishing *domains* for phishing and advanced *persistent* threat (APT) attacks, making it possible for security researchers to identify sites using natural language processing (NLP) techniques.

---

***New Words and Expressions***

**cybercriminal**/ˈsaɪbəˌkrɪmɪnl/ n.
　计算机罪犯
**lexical**/ˈleksɪk(ə)l/ adj.
　词汇的
**domain**/də(ʊ)ˈmeɪn/ n.
　领域；域名
**persistent**/pəˈsɪst(ə)nt/ adj.
　固执的，坚持的

Figure 5-6    Phish

That's according to OpenDNS Security Labs, which is *prototyping* a tool *dubbed* NLPRank to see if it can identify potentially *malicious* websites and phishing domains more quickly. Based on tests so far, the natural language processing tool could prove to be a "robust" method for defending against APTs, claimed OpenDNS security researcher Jeremiah O'Connor in a blog post.

Security researchers at OpenDNS recently analyzed DNS data associated with attacks carried out by the *cybercrime* group behind the Carbanak malware, which is believed to have stolen hundreds of millions of dollars from banks around the world in a sophisticated, multiyear APT campaign.

## APT Campaigns

To *penetrate* banks and various other financial *institutions*, these cybercriminals would typically target employees through phishing emails laced with malware, which, when installed on a system, would allow them to take complete control of the compromised computer. At that point, they would move *laterally* across the network to other more critical systems, gain access to administrative accounts, control ATMs and *siphon* out huge sums of money.

When comparing the malicious domains and *spoofing* techniques used in the Carbanak campaign with those used in other APTs like the Darkhotel cyber *espionage* campaign, OpenDNS observed they were constructed in a similar lexical fashion. "One of the spoofing techniques often leveraged is the impersonation of a legitimate software or tech company in an email claiming a required software update," O'Connor said.

*New Words and Expressions*

**spear**/spɪə/ n.

矛，枪

**tactic**/ˈtæktɪk/ n.

策略，战略

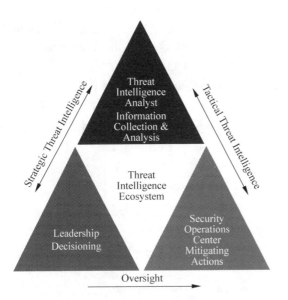

Figure 5-7    APT Attack

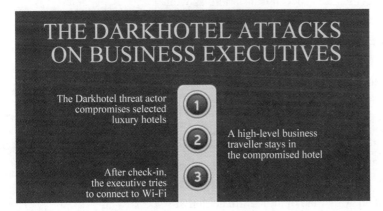

Figure 5-8    Darkhotel Attack

Domains used in the Darkhotel campaign, for example, included adobeupdates.com, adobeplugs.net, adoberegister. flashserv.net and microsoft-xpupdate.com. Meanwhile, the Carbanak APT used domains such as update-java.net and adobe-update.net. Other instances of domain names sharing a similar lexical structure included gmailboxes.com, microsoft-update-info.com and firefoxupdata.com.

**Lexical Similarities**

In reviewing the attack data, OpenDNS discovered multiple cases of suspicious websites advertising fake Java updates, sharing the same infrastructure and exhibiting similar attack patterns, O'Connor said. Researchers discovered that APT groups have a tendency to spoof legitimate domains and use *spear* phishing *tactics* to *obfuscate* their criminal campaigns.

Because of the lexical similarities among the domains used in these criminal campaigns, it is possible to use NLP techniques to identify potentially malicious *typo-squatting* and targeted phishing domains, O'Connor said. NLP is basically a technique for extracting meaning from written words using specialized software. Its tools are used widely to read and interpret free text documents in a variety of applications and fields.

### Natural Language Processing via Minimum-Edit Distance

According to O'Connor, OpenDNS' NLPRank system uses NLP, HTML tag analysis and a method known as minimum-edit distance to see if it can distinguish between legitimate and malicious domains on the Internet.

The minimum-edit distance method checks for the distance between words in legitimate and typo-squatting domains. It is used in other applications like spell-checking and speech translation, as well, and offers a way to define and differentiate the language used by malicious domains from the one used by legitimate domains, O'Connor said.

Another process OpenDNS uses in *conjunction* with NLP to identify malicious domains is autonomous systems number (ASN) mapping. Malicious domains are usually hosted on IP networks that are not associated with the domain they're attempting to spoof. For example, if a domain offering an Adobe update maps to an IP network that does not belong to Adobe, there is a good chance the domain is malicious. OpenDNS has built an ASN map of all legitimate domains on the Internet along with their appropriate ASNs, O'Connor said.

Using these methods, NLPRank has reportedly been able to spot several types of phishing attacks spoofing major companies such as Wells Fargo, Facebook, Dropbox and others.

**New Words and Expressions**

**obfuscate**/ˈɒbfʌskeɪt/ v.
使模糊；使迷乱

**typo-squatting**
误植域名

**conjunction**/kənˈdʒʌŋ(k)ʃ(ə)n/ n.
结合

## 参考译文 B

### 使用自然语言处理来识别恶意域

网络犯罪分子在建立网络钓鱼和高级持久威胁（APT）攻击域时，显然倾向于使用相同（或至少类似）的词汇风格，这使得安全研究人员有可能使用自然语言处理技术来识别站点。

OpenDNS 安全实验室正在开发一种名为 NLPRank 的工具，看看它能否更快地识别出潜在的恶意网站和钓鱼域名。OpenDNS 安全研究人员 Jeremiah O'Connor 在一篇博客文章中称，根据迄今为止的测试，自然语言处理工具可能被证明是一种抵御 APTs 的"健壮"方法。

OpenDNS 的安全研究人员最近分析了 DNS 数据，这些数据与卡班纳克（Carbanak）恶意软件背后的网络犯罪集团实施的攻击有关。据说，该恶意软件进行了一场复杂的、持续多年的 APT 攻击，从世界各地的银行窃取了数亿美元。

**APT 活动**

为了渗透银行和其他各种金融机构，这些网络犯罪分子通常会通过带有恶意软件的网络钓鱼电子邮件攻击员工。恶意软件安装在系统上后，就能让他们完全控制受损的电脑。到那时，他们将通过网络横向转移到其他更关键的系统，获得管理账户的访问权限，控制自动取款机，并吸走巨额资金。

在比较卡班纳克战役中使用的恶意域和欺骗技术与暗黑酒店网络间谍活动等其他 APTs 中使用的恶意域和欺骗技术时，OpenDNS 注意到它们是用类似的词汇方式构建的。O'Connor 说："欺骗技术的一个常用手段是在一封声称需要软件更新的电子邮件中冒充合法软件或技术公司。"

例如，"暗黑酒店"活动中使用的域名包括 adobeupdates.com、adobeplugs.net、adoberegister.flashserv.net 和 microsoft-xpupdate.com。同时，卡班纳克 APT 使用了 update-java.net 和 adobe-update.net 等域。其他共享类似词汇结构的域名实例包括 gmailboxes.com、microsoft-update-info.com 和 firefoxupdata.com。

**词汇的相似之处**

O'Connor 说，在审查攻击数据时，OpenDNS 发现了多起可疑网站发布虚假 Java 更新、共享相同基础设施和显示类似攻击模式的案例。研究人员发现，APT 群体有一种欺骗合法域名的倾向，他们使用鱼叉式网络钓鱼策略来混淆他们的犯罪活动。

O'Connor 接着说，由于这些犯罪活动中使用的域名在词汇上有相似之处，因此有可能使用自然语言处理技术来识别潜在的误植域名和有针对性的钓鱼域名。自然语言处理基本上是一种使用专门的软件从书面文字中提取意思的技术。它的工具被广泛用于阅读和解释各种应用程序和领域中的免费文本文档。

**通过最小编辑距离进行自然语言处理**

据 O'Connor 介绍，OpenDNS 的 NLPRank 系统使用自然语言处理、HTML 标签分析和一种被称为"最小编辑距离"的方法来区分互联网上的合法和恶意域名。

最小编辑距离方法检查合法域和字体占用域中的单词之间的距离。O'Connor 说，它还可用于拼写检查和语音翻译等其他应用程序，并提供了一种定义和区分恶意域名使用的语言和合法域名使用的语言的方法。

OpenDNS 与自然语言处理联合使用的另一个识别恶意域的过程是自动系统编号映射。恶意域名通常驻留在 IP 网络上，这些网络与恶意域名试图欺骗的域名无关。例如，如果提供 Adobe 更新的域映射到不属于 Adobe 的 IP 网络，则该域很可能是恶意的。O'Connor 说，OpenDNS 已经建立了一个 ASN 地图，包含了互联网上所有合法域名及其相应的 ASN。

据报道，使用这些方法，NLPRank 能够识别欺骗富国银行（Wells Fargo）、Facebook、Dropbox 等大公司的几种钓鱼攻击。

# Chapter *6*

# Artificial Intelligence in Agriculture

## Text A

According to UN Food and *Agriculture* Organization[1], the *population* will increase by 2 billion by 2050. However, only 4% *additional* land will come under *cultivation* by then.

In this context, use of latest *technological* solutions to make farming more efficient, remains one of the greatest *imperatives*. While Artificial Intelligence[2] (AI) sees a lot of direct application across sectors, it can also bring a *paradigm* shift in how we see farming today. AI-powered solutions will not only enable farmers to do more with less, it will also improve quality and ensure faster go-to-market for crops.

In this article, we will discuss how AI can change the agriculture *landscape*, the application of drone-based image processing techniques, precision farming landscape, the future of agriculture and the challenges ahead.

### Scope of AI in Agriculture

Agriculture is seeing rapid adoption of Artificial Intelligence (AI) and Machine Learning[3] (ML) both in terms of agricultural products and in-field farming techniques. Cognitive computing in particular is all set to become the most *disruptive* technology in agriculture services as it can understand, learn, and respond to different *situations* (based on learning) to increase *efficiency*.

Providing some of these solutions as a service like chatbot[4] or other conversational *platform* to all the farmers will help them keep

---

**New Words and Expressions**

**agriculture**/ˈæɡrɪkʌltʃə/ n.
农业；农耕；农业生产；农学
**population**/pɒpjʊˈleɪʃ(ə)n/ n.
人口；【生物】种群，【生物】群体；全体居民
**additional**/əˈdɪʃ(ə)n(ə)l/ adj.
附加的，额外的
**cultivation** /kʌltɪˈveɪʃn/ n.
培养；耕作；耕种；教化；文雅
**technological**
/teknəˈlɒdʒɪk(ə)l/ adj.
技术的；工艺的
**imperatives**/ɪmˈperətɪvz/ n.
祈使句；命令句
**paradigm**/ˈpærədaɪm/ n.
范例；词形变化表
**landscape**/ˈlændskeɪp/ n.
风景；风景画；景色；山水画；乡村风景画；地形
**disruptive**/dɪsˈrʌptɪv/ adj.
破坏的；分裂性的；制造混乱的
**situations**/ˌsɪtʃʊˈeɪʃnz/ n.
状况；情境；局面
**efficiency**/ɪˈfɪʃ(ə)nsɪ/ n.
效率；效能；功效
**platform**/ˈplætfɔːm/ n.
平台；月台，站台；坛；讲台

pace with technological *advancements* as well as apply the same in their daily farming to reap the benefits of this service.

Currently, Microsoft is working with 175 farmers in Andhra Pradesh, India to provide *advisory* services for sowing, land, *fertilizer* and so on. This initiative has already resulted in 30% higher yield per *hectare* on an average compared to last year.

Given below are top five areas where the use of *cognitive* solutions can benefit agriculture.

## 1. Growth Driven by IOT[5]

Huge volumes of data get generated every day in both structured and unstructured format. These relate to data on historical weather pattern, soil reports, new research, rainfall, pest infestation, images from Drones and cameras and so on. Cognitive IOT solutions can sense all this data and provide strong insights to improve yield.

Proximity Sensing[6] and Remote Sensing[7] are two technologies which are *primarily* used for intelligent data *fusion*. One use case of this high-resolution data is Soil Testing[8]. While remote sensing requires sensors to be built into *airborne* or *satellite* systems, *proximity* sensing requires sensors in contact with soil or at a very close range. This helps in soil characterization based on the soil below the surface in a particular place.

Hardware solutions like Rowbot (pertaining to corns) are already pairing data-collecting software with robotics to prepare the best fertilizer for growing f corns in addition to other activities to maximize output.

## 2. Image-Based Insight Generation

*Precision* farming is one of the most discussed areas in farming today. Drone-based images can help in in-depth field analysis, crop *monitoring*, scanning of fields and so on.

Computer vision technology, IOT and drone data can be combined to ensure rapid actions by farmers. Feeds from drone image data can generate alerts in real time to accelerate precision farming

Companies like Aerialtronics have *implemented* IBM Watson IoT Platform and the Visual Recognition APIs in *commercial drones* for image analysis in real time. Given below are some areas where computer vision technology can be put to use:

---

### New Words and Expressions

**advancement**/əd'vɑːnsm(ə)nt/ n.
前进，进步；提升

**advisory**/əd'vaɪz(ə)rɪ/ adj.
咨询的；顾问的；劝告的

**fertilizer**/'fɜːtɪlaɪzə/ n.
肥料；受精媒介物；促进发展者

**hectare**/'hekteə; -ɑː/ n.
公顷

**cognitive**/'kɒgnɪtɪv/ adj.
认知的，认识的

**primarily**/'praɪm(ə)rɪlɪ/ adv.
首先；主要地，根本上

**fusion**/'fjuːʒ(ə)n/ n.
融合；熔化，熔接；融合物

**airborne**/'eəbɔːn/ adj.
【航】空运的；空气传播的；风媒的

**satellite**/'sætəlaɪt/ n.
卫星；人造卫星；随从；卫星国家

**proximity**/prɒk'sɪmɪtɪ/ n.
接近，【数】邻近；接近；接近度，距离；亲近

**precision**/prɪ'sɪʒ(ə)n/ adj.
精密的，精确的

**monitoring**/'mɒnɪtərɪŋ/ n.
监视，【自】监控；检验，检查

**implemented**/'ɪmplɪm(ə)nted/ v.
实施，执行

**commercial**/kə'mɜːʃ(ə)l/ adj.
商业的；赢利的；靠广告收入的

**drone**/drəʊn/ n.
雄蜂；嗡嗡的声音；懒惰者

- Disease detection:

Figure 6-1  disease detection

Preprocessing of image ensure the leaf images are segmented into areas like background, non-diseased part and diseased part. The diseased part is then cropped and send to remote labs for further *diagnosis*. It also helps in pest *identification*, nutrient deficiency recognition and more.

- Crop *readiness* identification:

Images of different crops under white/UV-A light are captured to determine how ripe the green fruits are. Farmers can create different levels of readiness based on the crop/fruit *category* and add them into separate stacks before sending them to the market.

- Field management:

Using high-definition images from airborne systems (drone or copters), real-time estimates can be made during cultivation period by creating a field map and identifying areas where crops require water, fertilizer or *pesticides*. This helps in resource *optimization* to a huge extent.

### 3. Identification of Optimal Mix for *Agronomic* Products

Based on *multiple* parameters like soil condition, weather forecast, type of seeds, *infestation* in a certain area and so on, cognitive solutions make recommendations to farmers on the best choice of crops and hybrid seeds. The *recommendation* can be further personalized based on the farm's requirement, local conditions, and data about successful farming in the past. External factors like *marketplace* trends, prices or consumer needs may also be factored into enable farmers take a well-informed decision.

### 4. Health Monitoring of Crops

Remote sensing techniques along with *hyper* spectral imaging[9] and 3D laser scanning[10] are essential to build crop metrics across thousands of acres. It has the potential to bring in a *revolutionary* change in terms of how farmlands are monitored by farmers both from time and effort *perspective*. This technology will also be used to monitor crops along their entire lifecycle including report generation in case of anomalies.

**New Words and Expressions**

**diagnosis**/ˌdaɪəɡˈnəʊsɪs/ n.
诊断

**identification**/aɪˌdentɪfɪˈkeɪʃ(ə)n/ n.
鉴定，识别；认同；身份证明

**readiness**/ˈredɪnəs/ n.
敏捷，迅速；准备就绪；愿意

**category**/ˈkætɪɡ(ə)rɪ/ n.
种类，分类；【数】范畴

**pesticides**/ˈpestɪsaɪdz/ n.
农药；杀虫剂

**optimization**
/ˌɒptɪmaɪˈzeɪʃən/ n.
最佳化，最优化

**agronomic**/ˌæɡrəˈnɒmɪk/ adj.
农事的；农艺学的

**multiple**/ˈmʌltɪpl/ adj.
多重的；多样的；许多的

**infestation**/ˌɪnfeˈsteɪʃən/ n.
感染；侵扰

**recommendation**
/ˌrekəmenˈdeɪʃ(ə)n/ n.
推荐；建议；推荐信

**marketplace**/ˈmɑːkɪtpleɪs/ n.
市场；商场；市集

**hyper**/ˈhaɪpə/ adj.
亢奋的；高度紧张的

**revolutionary**/revəˈluːʃ(ə)n(ə)rɪ/ adj.
革命的；旋转的；大变革的

**perspective**/pəˈspektɪv/ n.
观点；远景；透视图

## 5. Automation Techniques in Irrigation and Enabling Farmers

In terms of human *intensive* processes in farming, *irrigation* is one such process. Machines trained on historical weather pattern, soil quality and kind of crops to be grown, can automate irrigation[11] and increase overall yield. With close to 70% of the world's fresh water being used in irrigation, automation can help farmers better manage their water problems.

### Importance of Drone

As per a recent PWC Study, the total addressable market for Drone-based solutions across the globe is $127.3 billion and for agriculture it is at $32.4 billion.

Drone-based solutions in agriculture have a lot of significance in terms of managing adverse weather conditions, productivity gains, precision farming and yield management.

Before the crop cycle, drone[12] can be used to produce a 3D field map of detailed terrain, *drainage*, soil viability and irrigation. Nitrogen-level management can also be done by drone solutions.

Aerial spraying of pods with seeds and plant nutrients into the soil provides necessary supplements for plants. Apart from that, Drones can be programmed to spray liquids by modulating distance from the ground depending on the terrain.

Crop Monitoring[13] and Health assessment[14] remains one of the most significant areas in agriculture to provide drone-based solutions in collaboration with Artificial Intelligence and computer vision technology. High-resolution cameras indrones collect precision field images which can be passed through convolution neural network to identify areas with weeds, which crops need water, plant stress level in mid-growth stage. In terms of *infected* plants, by scanning crops in both RGB and near-infra red light, it is possible to generate *multispectral* images using drone devices. With this, it is possible to specify which plants have been infected including their location in a vast field to apply remedies, *instantly*. The multi *spectral* images combinehyper spectral images with 3D scanning techniques to define the spatial information system that is used for acres of land. The *temporal* component provides the guidance for the entire lifecycle of the plant.

---

### New Words and Expressions

**intensive**/ɪnˈtensɪv/ adj.
加强的；集中的；透彻的；加强语气的

**irrigation**/ˌɪrɪˈɡeɪʃn/ n.
灌溉；[临床]冲洗；冲洗法

**drainage**/ˈdreɪnɪdʒ/ n.
排水；排水系统；污水；排水面积

**infected**/ɪnˈfektɪd/ adj.
被感染的

**multispectral**/mʌltɪˈspektr(ə)l/ adj.
多谱线的

**instantly**/ˈɪnst(ə)ntlɪ/ adv.
立即地；马上地；即刻地

**spectral**/ˈspektr(ə)l/ adj.
光谱的；幽灵的；鬼怪的

**temporal**/ˈtemp(ə)r(ə)l/ adj.
暂时的；当时的；现世的

## Precision Farming

The phrase "Right Place, Right Time, Right Product" sums up precision farming. This is a more *accurate* and controlled technique that replaces the *repetitive* and labor-intensive part of farming. It also provides guidance about crop *rotation*, *optimum* planting and harvesting time, water management, *nutrient* management, pest attacks and so on.

Key technologies that enable precision farming are given below:

- High precision positioning system
- Automated steering system
- Geo mapping
- Sensor and remote sensing
- Integrated electronic communication
- *Variable* rate technology

Goals for precision farming:

- Profitability: Identifying crops and market *strategically* as well as predicting ROI based on cost and margin.
- Efficiency: By investing in precision algorithm, better, faster and cheaper farming opportunities can be utilized. This enables overall accuracy and efficient use of resource
- Sustainability: Improved social, environmental and economic performance ensures *incremental* improvements each season for all the performance indicators

Examples of precision farming management:

- Identification of stress level in a plant is obtained from high-resolution images and multiple sensor data on plants.

This large set of data from multiple sources needs to be used as an input for Machine Learning to enable data fusion and feature identification for stress recognition.

- Machine learning models trained on plant images can be used to recognize stress levels in plants. The entire approach can be classified into four stages of identification, classification, quantification and prediction to make better decisions.

## Yield Management Using AI

The *emergence* of new age technologies like Artificial Intelligence (AI), Cloud Machine Learning, Satellite Imagery and advanced analytics are creating an *ecosystem* for smart farming. Fusion of all this technology is enabling farmers achieve higher average yield

### New Words and Expressions

**accurate**/ˈækjərət/ adj.
精确的

**repetitive**/rɪˈpetɪtɪv/ adj.
重复的

**rotation**/rə(ʊ)ˈteɪʃ(ə)n/ n.
旋转；循环，轮流

**optimum**/ˈɒptɪməm/ adj.
最适宜的

**nutrient**/ˈnjuːtrɪənt/ adj.
营养的；滋养的

**variable**/ˈveərɪəb(ə)l/ adj.
变量的；可变的；易变的，多变的

**strategically**/strəˈtiːdʒɪkəlɪ/ adv.
战略性地；战略上

**algorithm**/ˈælgərɪð(ə)m/ n.
算法，运算法则

**incremental**/ˌɪnkrɪˈmentəl/ adj.
增加的，增值的

**emergence**/ɪˈmɜːdʒ(ə)ns/ n.
出现，浮现；发生；露头

**ecosystem**/ˈiːkəʊsɪstəm/ n.
生态系统

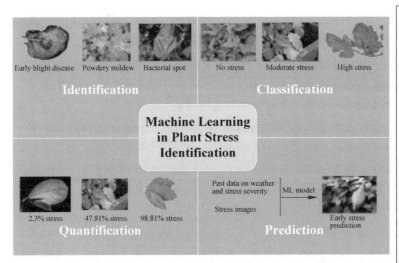

Figure 6-2　machine learning in plant stress identification

*New Words and Expressions*

**preparation**/ˌprepəˈreɪʃ(ə)n/ n.

预备；准备

**fertilization**/ˌfɜːtɪlaɪˈzeɪʃ(ə)n/ n.

施肥；【胚】受精；肥沃

**optimum**/ˈɒptɪməm/adj.

最适宜的

**advisory**/ədˈvaɪz(ə)rɪ/ adj.

咨询的；顾问的；劝告的

**climate**/ˈklaɪmɪt/ n.

气候；风气；思潮；风土

**rainfall**/ˈreɪnfɔːl/ n.

降雨；降雨量

**leverages**/ˈliːv(ə)rɪdʒ/ n.

手段，影响力；杠杆作用；杠杆效率

and better price control.

Microsoft is currently working with farmers from Andhra Pradesh to provide advisory services using Cortana Intelligence Suite including Machine Learning and Power BI. The pilot project uses an AI sowing app to recommend sowing date, land *preparation*, soil test-based *fertilization*, farm yard manure application, seed treatment, *optimum* sowing depth and more to farmers which has resulted in 30% increase in average crop yield per hectare.

Technology can also be used to identify optimal sowing period, historic *climate* data, real time Moisture Adequacy Data (MAI) from daily *rainfall* and soil moisture to build predictability and provide inputs to farmers on ideal sowing time.

To identify potential pest attacks, Microsoft in collaboration with United Phosphorus Limited is building a Pest Risk Prediction API that *leverages* AI and machine learning to indicate in advance, the risk of pest attack. Based on the weather condition and crop growth stage, pest attacks are predicted as High, Medium or Low.

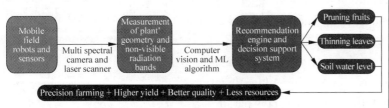

Figure 6-3　robotics helping in Digital Farming

AI *Startups* in Agriculture:

### 1. Prospera, founded in 2014.

This Israeli startup has revolutionized the way farming is done. It has developed a cloud-based solution that aggregates all existing data that farmers have like soil/water sensors, aerial images and so on. It then combines it with an in-field device that makes sense of it all. The Prospera device which can be used in green houses or in the field is powered by a variety of sensors and technologies like computer vision. The inputs from these sensors are used to find a correlation between different data labels and make predictions.

### 2. Blue River technology, founded in 2011.

This California-based startup combines artificial intelligence, computer vision and robotics to build next-generation agriculture equipment that reduces *chemicals* and saves costs. Computer vision identifies each individual plant, ML decides how to treat each individual plant and robotics enables the smart machines to take action.

### 3. FarmBot, founded in 2011.

This company has taken precision farming to a different level by enabling environment conscious people with precision farming technology to grow crops at their own place. The product, FarmBot comes at a price of $4000 and helps the owner to do end-to-end farming all by himself. Ranging from seed plantation to weed detection and soil testing to watering of plants, everything is taken care of by this physical bot using an open source software system.

## Challenges in AI Adoption in Agriculture

Though Artificial Intelligence others vast opportunities for application in agriculture, there still exists a lack of familiarity with high tech machine learning solutions in farms across most parts of the world. *Exposure* of farming to external factors like weather conditions, soil conditions and presence of pests is quite a lot. So what might look like a good solution while planning during the start of *harvesting*, may not be an optimal one because of changes in external parameters.

AI systems also need a lot of data to train machines and to make precise predictions. In case of vast agricultural land, though spatial data can be gathered easily, temporal data is hard to get. For example, most of the crop-specific data can be obtained only once in a year when the crops are growing. Since the data infrastructure

| **New Words and Expressions** |
| --- |

**startup**/stɑːtʌp/ n.
启动；开办
**chemical**/ˈkemɪk(ə)l/ n.
化学制品，化学药品
**exposure**/ɪkˈspəʊʒə; ek-/ n.
暴露；曝光；揭露；陈列
**harvesting**/ˈhɑːvɪstɪŋ/ n.
收割；收获

takes time to mature, it requires a significant amount of time to build a robust machine learning model.

This is one reason why AI sees a lot of use in *agronomic* products such as seeds, fertilizer, pesticides and so on rather than in-field precision solutions.

## Summary

The future of farming depends largely on adoption of *cognitive* solutions. While large scale research is still in progress and some applications are already available in the market, the industry is still highly underserved. When it comes to handling realistic challenges faced by farmers and using autonomous decision making and predictive solutions to solve them, farming is still at a *nascent* stage.

In order to explore the enormous scope of AI in agriculture, applications need to be more robust. Only then will it be able to handle frequent changes in external conditions, *facilitate* real-time decision making and make use of appropriate framework/platform for collecting *contextual* data in an efficient manner.

Another important aspect is the *exorbitant* cost of different cognitive solutions available in the market for farming. The solutions need to become more affordable to ensure the technology reaches the masses. An open source platform would make the solutions more affordable, resulting in rapid adoption and higher *penetration* among the farmers.

---

**New Words and Expressions**

**agronomic**/ˌægrəˈnɒmɪk/ adj.
　农事的；农艺学的

**cognitive**/ˈkɒgnɪtɪv/ adj.
　认知的，认识的

**nascent**/ˈnæs(ə)nt / adj.
　初期的；开始存在的，发生中的

**facilitate**/fəˈsɪlɪteɪt/ v.
　促进；帮助；使容易

**contextual**/kɒnˈtekstjʊəl/ adj.
　上下文的；前后关系的

**exorbitant**/ɪgˈzɔːbɪt(ə)nt/ adj.
　过高的；（性格等）过分的

**penetration**/penɪˈtreɪʃ(ə)n/ n.
　渗透；突破；侵入；洞察力

---

## Terms

### 1. UN Food and Agriculture Organization

联合国粮食农业组织是联合国系统内最早的常设专门机构。其宗旨是提高人民的营养水平和生活标准，改进农产品的生产和分配，改善农村和农民的经济状况，促进世界经济的发展并保证人类免于饥饿。它是各成员国间讨论粮食和农业问题的国际组织。

### 2. Artificial Intelligence

人工智能是研究、开发用于模拟、延伸和扩展人的智能的理论、方法、技术及应用系统的一门新的技术科学。

### 3. Machine Learning

机器学习是一门多领域交叉学科，涉及概率论、统计学、逼近论、凸分析、算法复杂度理论等多门学科。专门研究计算机怎样模拟或实现人类的学习行为，以获取新的知识或技能，重新组织已有的知识结构使之不断改善自身的性能。

### 4. Chatbot

聊天机器人（Chatbot）可用于实用的目的，如客户服务或信息获取。有些聊天机器人会搭载自然语言处理系统，但大多简单的系统只会撷取输入的关键字，再从数据库中找寻最合适的应答句。目前，聊天机器人是虚拟助理（如 Google 智能助理）的一部分，可以与许多组织的应用程序、网站以及即时消息平台（Facebook Messenger）连接。非助理应用程序包括娱乐目的的聊天室、研究特定产品的促销、社交机器人。

**5. IOT**

物联网（IOT）是新一代信息技术的重要组成部分，也是"信息化"时代的重要发展阶段。物联网就是物物相连的互联网。

**6. Proximity Sensing**

接近报警（Proximity Sensing）是一种当入侵者接近它（但还未碰到它）时能触发报警的探测装置。

**7. Remote Sensing**

遥感（Remote Sensing）是指非接触的、远距离的探测技术。一般指运用传感器/遥感器对物体的电磁波的辐射、反射特性的探测。遥感是通过遥感器这类对电磁波敏感的仪器，在远离目标和非接触目标物体条件下探测目标地物。

**8. Soil Testing**

土壤测试（Soil Testing）是指经田间试验证明测定结果与作物产量之间有最佳相关性的分析方式。

**9. Hyper Spectral Imaging**

高光谱成像技术（Hyper Spectral Imaging）是基于非常多窄波段的影像数据技术，它将成像技术与光谱技术相结合，探测目标的二维几何空间及一维光谱信息，获取高光谱分辨率的连续、窄波段的图像数据。

**10. 3D Laser Scanning**

它是测绘领域继 GPS 技术之后的一次技术革命。它突破了传统的单点测量方法，具有高效率、高精度的独特优势。三维激光扫描技术能够提供扫描物体表面的三维点云数据，因此可以用于获取高精度高分辨率的数字地形模型。

**11. Automate Irrigation**

灌溉自动化。农业是用水大户，近年来农业用水量约占经济社会用水总量的 62%，部分地区高达 90%以上，农业用水效率不高，节水潜力很大。大力发展农业节水，在农业用水量基本稳定的同时扩大灌溉面积、提高灌溉保证率，是促进水资源可持续利用、保障国家粮食安全、加快转变经济发展方式的重要举措。

**12. Drone**

它是利用无线电遥控设备和自备的程序控制装置操纵的无人飞机，或者由车载计算机完全地或间歇地自主地操作。

**13. Crop Monitorin**

农作物检测，以仪器或感官鉴别作物品质的操作程序和方法。检验结果按统一标准分级后，可据以确定作物在生产上的使用价值，并为贮藏、运输、推广销售和引种交换等提供科学依据。

**14. Health Assessment**

它是指通过涉及健康的危险性因素分析，得出影响健康的综合因素的评价报告。健康管理者及个人能够清楚地了解个人健康状态。

## Comprehension

### Blank Filling

1. _____ is all set to become the most disruptive technology in agriculture services as it can _____ ,_____ , and _____to different situations to increase efficiency.

2. Drone-based solutions in agriculture have a lot of significance in terms of managing _____, _____, _____ and _____.

3. _____ and _____ remains one of the most significant areas in agriculture to provide drone-based solutions in collaboration with _____ and _____.

4. Machine learning models trained on _____ can be used to recognize stress levels in plants.

5. Machine learning models can be classified into four stages of _____, _____, _____and _____.

### Content Questions

1. What are the five areas where using cognitive solutions is good for agriculture?

2. What is the definition of artificial intelligence?

3. What is the status quo of artificial intelligence industry?

4. What are the key technologies to achieve precision agriculture?

5. What is the target of precision agriculture?

## Answers

### Blank Filling

1. Cognitive computing; understand; learn; respond

2. adverse weather conditions; productivity gains; precision farming; yield management

3. Crop Monitoring; Health assessment; Artificial Intelligence; computer vision technology

4. plant images

5. identification; classification; quantification; prediction

### Content Questions

1. Growth driven by IOT、Image-based insight generation、Identification of optimal mix for agronomic products、Health monitoring of crops、Automation techniques in irrigation and enabling farmers

2. It is a new technology science to research and develop theories,methods, technologies and application systems for simulating, extending and extending human intelligence.

3. The industrial chain of artificial intelligence can be divided into infrastructure layer, application technology layer and industry application layer.

4. High precision positioning system、Automated steering system、Geo mapping、Sensor and remote sensing、 Integrated electronic communication、 Variable rate technology.

5. Profitability、 Efficiency、 Sustainability.

## 参考译文 A

根据联合国粮农组织的数据，到 2050 年，世界人口将增加 20 亿，但那时只有 4%的新增土地可用于耕种。

在这方面，利用最新技术提高农业效率仍然是当务之急之一。人工智能(AI)在各个领域都有很多直接应用，它也能改变传统农业。人工智能解决方案不仅能让农民花更少的钱做更多的事，还能提高质量，确保更快地将作物投放市场。

在本文中，我们将讨论人工智能如何改变农业，基于无人机的图像处理技术的应用，精准农业，农业的未来和挑战。

### 人工智能在农业中的应用范围

采用人工智能（AI）和机器学习（ML）的田间耕作技术在农业领域迅速发展。特别是认知计算，它将成为农业服务领域最具颠覆性的技术，因为它可以理解、学习和响应不同的情景来提高效率。

将这些解决方案中的一部分以聊天机器人或其他通信的形式反馈给农业从业者，将有助于他们跟上技术进步的步伐，并将其应用于日常农业，从而获得更好的服务。

目前，微软正在与印度安得拉邦的 175 名农民合作，为他们提供播种、土地、化肥等方面的咨询服务。这项措施已经使每公顷土地的平均产量比去年提高了 30%。

以下是使用认知解决方案对农业有益的五大领域。

### 1. 物联网驱动的增长

物联网每天都会生成大量结构化和非结构化数据。这些数据与历史气候模式、土壤报告、新研究、降雨、虫害、无人机和照相机拍摄的图像等有关。认知物联网解决方案可以感知所有这些数据，并对如何提高产量进行剖析。

近程感知和遥感是两种主要用于智能数据融合的技术。这种高精度数据可以用于土壤测试。虽然遥感需要将传感器植入机载或卫星系统，但近距离遥感需要传感器与土壤接触或在非常近的距离才有效。这有助于描述特定地区地表以下的土壤特征。

像 Rowbot（与玉米有关）这样的硬件解决方案已经将数据收集软件与机器人技术结合起来，为种植玉米准备最好的肥料，同时还进行其他活动，以最大限度地提高产量。

### 2. 基于图像的洞察力

精耕细作是当今农业中讨论最多的领域之一。基于无人机的图像可以帮助进行深入的田间分析、作物监测和田间扫描等。

计算机视觉技术、物联网和无人机数据可以结合起来，确保农民迅速采取行动。来自无人机图像数据的反馈可以实时生成警报，加速精准农业。

Aerialtronics 等已经将商用无人机连接到了 IBM Watson 物联网平台，可以实时进行视觉识别并分析图像。以下是一些可应用电脑视觉技术的范畴。

- 植物病害检查

图像预处理把树叶图像分成背景、非病变区域和病变区域，然后对病变区域进行裁剪，送到远程实验室进行进一步诊断。它还有助于对害虫的识别、植物营养元素缺乏的判断等。

- 作物成熟度识别

在白光/长波紫外线的照射下,用不同作物的图像来确定绿色果实的成熟程度。农民可以根据作物/水果类别创建不同的级别,并将它们进行分类,然后再发送到市场。

- 现场管理

利用空中系统(无人机或直升机)提供的高清晰图像,可以在种植期间通过创建农田地图和确定作物需要水、化肥或杀虫剂的地区来进行实时估计。这在很大程度上有助于资源优化。

**3. 确定农艺产品的最佳组合**

基于土壤条件、天气预报、种子类型、某一地区的虫害等多个参数,可以为农民进行农作物的种植推荐,还推荐可以根据农场的需求、当地条件和过去成功种植的数据进行个性化推荐。市场趋势、价格或消费者需求等外部因素也可能被考虑在其中,来帮助农户做出决定。

**4. 农作物健康监察**

遥感技术以及超光谱成像和三维激光扫描通过对农田的监测,有可能给农户的耕种时间和耕种行为带来巨大的变化。这项技术还将对农作物的整个生长周期进行监测,并对监测到的异常情况进行及时的报告。

**5. 灌溉自动化技术和支持农民**

在农业生产过程中,可以通过集约灌溉节约水资源。机器根据历史气候模式、土壤质量和将要种植的作物种类,可以实现灌溉自动化,提高整体产量。由于全球近70%的淡水用于灌溉,自动化可以帮助农民更好地解决水资源问题。

**无人驾驶的重要性**

普华永道(PWC)最近的一项研究显示,全球无人机解决方案的潜在市场总额为1273亿美元,农业市场为324亿美元。

基于无人机的农业解决方案在应对恶劣天气、提高生产力、精耕细作和产量管理等方面具有重要意义。

在耕种之前,无人机可以用来制作详细的地形、排水、土壤活力和灌溉的三维野外地图。氮水平管理也可以通过无人机解决方案来实现。

向土壤中喷撒豆荚种子和植物养分,为植物提供必要的补充。除此之外,无人机还可以根据地形调整与地面的距离,从而设定喷淋液体的程序。

无人机作物监测和健康评估是人工智能和计算机视觉技术在农业领域最重要的应用。无人机上的高分辨率相机可以采集精确的野外图像,通过卷积神经网络对杂草生长的区域进行识别,确定哪些作物需要水分,哪些作物处于生长中期水平。对于受感染的植物,利用 RGB 和近红外光扫描作物,可以利用无人机设备生成多光谱图像。有了这一技术,就有可能确定哪些植物已经被感染,并确定它们的具体位置,以便立即采取补救措施。多光谱图像将超光谱图像与三维扫描技术相结合,是为大面积的土地服务的空间信息系统。时态组件为植物的整个生命周期提供了指导。

**精准农业**

"对的地方,对的时间,对的产品"这句话概括了精耕细作。这是一种更为精确更可控的技术,取代了农业中重复性和劳动密集型的部分。它还提供了关于作物轮作、最佳种植和收获时间、灌溉、营养管理、虫害防治等方面的指导。

实现精准农业的关键技术如下：

- 高精度定位系统
- 自动转向系统
- 地理制图
- 传感器和遥感
- 集成电子通信
- 可变费率技术

精准农业目标：

- 盈利能力：从战略上识别作物和市场，并根据成本和利润率预测投资回报率。
- 效率：通过精准的数据，可以使农业更好更快地发展，降低成本。这样可以全面准确和有效地使用资源。
- 可持续性：社会、环境和经济业绩的改善确保了所有业绩指标在每个季度都有改善。

精耕细作管理的例子：

- 植物的应力水平识别是通过高分辨率图像和植物上的多个传感器数据获得的。

将多来源的数据输入进行机器学习，以便进行数据融合和特征识别，进行应力识别。

- 基于植物图像的机器学习模型可以用来识别植物的应激水平。整个方法可以分为识别、分类、量化和预测四个阶段，以做出更好的决策。

### 人工智能收益管理

人工智能（AI）、云机器学习、卫星图像和先进分析等新时代技术的出现，正在为智能农业创造一个生态系统。所有这些技术的融合使农产品的产量提高，同时也更好地控制成本。

微软目前正与来自安得拉邦的农民合作，使用 Cortana 智能套件（包括机器学习和 Power BI 决策分析系统）提供咨询服务。该试点项目利用人工智能播种 App，向农户推荐播种日期、土地整理、土壤试验施肥、农家肥施用、种子处理、最佳播种深度等，平均产量提高了 30%。

技术还可以用来确定最佳播种期、历史气候数据、来自每日降雨和土壤湿度的实时数据（MAI），从而进行预测，并为农民提供理想播种时间的建议。

为了识别潜在的病虫害，微软与联合磷有限公司合作，正在构建一个害虫风险预测 API，利用人工智能和机器学习提前指出病虫害的风险。根据气候条件和作物生长阶段，对害虫的危害程度进行预测，分为高、中、低三种。

### 农业 AI 创业

#### 1. 普洛斯佩拉，成立于 2014 年

这家以色列创业公司彻底改变了农业生产方式。它开发了一种基于云计算的解决方案，可以聚合农民现有的所有数据，如土壤/水传感器、航空图像等。然后，它将其与现场设备结合起来，使其具有意义。普洛斯佩拉的这种装置可以在温室里使用，也可以在野外使用，它由各种传感器和计算机视觉等技术组成。这些传感器的输入被用来发现不同数据标签之间的相关性，并做出预测。

#### 2. 蓝河科技，成立于 2011 年

这家总部位于加州的初创公司将人工智能、计算机视觉和机器人技术结合在一起，制造出下一代农业设备，这种设备可以减少化学材料并节约成本。计算机视觉识别每个植物个体，ML 决定如何处理每个植物个体，机器人技术使智能机器能够采取对应的措施。

**3. 开心农场机器人成立于 2011 年**

这家公司将精准农业提升到了一个新的高度，让拥有精准农业技术的环保人士可以在自己的土地上种植农作物。这款名为 FarmBot 的产品售价为 4000 美元，可以帮助农场主自己完成端对端耕作。从种子种植到杂草检测和土壤测试到植物浇水，一切都由这个使用开源软件系统的物理机器人来处理。

**农业人工智能应用的挑战**

虽然人工智能为农业应用提供了巨大的机会，但世界上大部分地区的农场仍然缺乏对高科技机器学习解决方案的了解。农业暴露于天气条件、土壤条件和害虫等外部因素的影响是相当大的。因此，在收割初期进行规划时，看起来可能是一个好的解决方案，但由于外部参数的变化，可能不是最佳方案。

人工智能系统还需要大量的数据来训练机器和做出精确的预测。在农业用地广阔的情况下，虽然空间数据很容易采集，但是时间数据很难获取。例如，大多数特定于作物的数据在作物生长时一年只能获得一次。由于数据基础设施需要时间来成熟，因此需要大量时间来构建健壮的机器学习模型。

这也是为什么人工智能在种子、化肥、农药等农艺产品中得到大量应用，而不是在现场精确解决方案中得到应用的原因之一。

**总结**

农业的未来在很大程度上取决于认知解决方案的采用。虽然大规模的研究仍在进行中，但是一些应用程序已经在市场上出现，但该行业仍然严重缺乏服务。在应对农民面临的现实挑战、运用自主决策和预测解决方案等方面，智能农业还处于起步阶段。

为了探索人工智能在农业中的巨大应用范围，应用程序需要更加健壮。只有这样才能够应对外部条件的频繁变化，方便实时决策，利用合适的框架/平台高效地收集数据。

另一个重要方面，是农业市场上不同的认知解决方案成本过高。解决方案需要变得更实惠，以确保技术普及到大众。开源平台将使这些解决方案变得更加经济实惠，从而在农民中迅速采用，获得更广泛的普及率。

# Text B

*Agriculture* and AI—*intuitively*, these domains seem to denote two separate worlds. Agriculture processes the soil and cares about food production and elementary supply—this is really down-to-earth! AI, in contrast, is deeply interwoven with computerized systems, complex interactions, modeling and reasoning approaches, and in public perception still suffers from flair of science fiction. So why do we dedicate the present special issue to this *combination*?

Modern Agriculture faces *tremendous* challenges. Today, the agricultural sector has grown into a highly competitive and *globalized* industry, where farmers and other actors have to consider local

---

***New Words and Expressions***

**agriculture**/ˈæɡrɪkʌltʃə/ n.
　农业；农艺，农学
**intuitively**/ɪnˈtjuːɪtɪvli/ adv.
　直观地；直觉地
**combination**/kɒmbɪˈneɪʃ(ə)n/ n.
　结合；组合；联合；[化学] 化合
**tremendous**/trɪˈmendəs/ adj.
　极大的，巨大的；惊人的；极好的
**globalize**/ˈɡləʊbəlaɪz/ v.
　使全球化

climatic and *geographic* aspects as well as global ecological and political factors in order to guarantee economic survival and *sustainable* production. Feeding a growing world population asks for continuous increases in food production, but arable land remains a limited resource. New requests for bio energy or changing diet preferences put *additional* strains on agricultural production, while *settlement* and transport consume increasing shares of land. Expected and observable changes in global climate, shifting rainfall patterns, global warming, *droughts*, or the increasing *frequency* and duration of extreme weather events endanger traditional production areas and bring new risks and uncertainties for global harvest yields. To cope with these challenges, Agriculture requires a continuous and sustainable increase in productivity and efficiency on all levels of agricultural production, while resources like water, energy, fertilizers etc. need to be used carefully and efficiently in order to protect and sustain the environment and the soil quality of the arable land. The complexity of the challenge is increased by other short-term events which are difficult to predict, such as epidemics, financial crisis, or price volatility for agricultural raw materials and products.

Consequently, Agriculture needs help in handling the *complexity*, uncertainty and fuzziness inherent in this domain, and it requires new solutions for all aspects of agricultural production-from better and predictable crop planning, to precision farming, optimized resource application, support of efficient and collaborative processes using modern technology, fully or partially *autonomous* solutions for tedious work, up to the sustained long-term development of useful knowledge resources.

The basis for computerized answers to such challenges in Agriculture has been realized in the recent decades: GPS (providing precise location data and offering the basis of all kinds of location-specific support) and mobile *communication* (allowing for the quick exchange of data between participants even in the field) are crucial and well-accepted breakthrough technologies. Making sense from the data that become available now, and using the resulting knowledge for process and operation improvement on all levels, brings into play AI and their modeling and reasoning capabilities.

From the AI point of view, Agriculture offers a vast application

---

### New Words and Expressions

**geographic**/ˌdʒɪəˈɡræfɪk/ adj.
地理的；地理学的

**sustainable**/səˈsteɪnəb(ə)l/ adj.
可以忍受的；足可支撑的；养得起的；可持续的

**additional**/əˈdɪʃ(ə)n(ə)l/ adj.
附加的，额外的

**settlement**/ˈset(ə)lm(ə)nt/ n.
定居，解决，处理

**drought**/draʊt/ n.
干旱；缺乏

**frequency**/ˈfriːkw(ə)nsɪ/ n.
频率；频繁

**complexity**/kəmˈpleksətɪ/ n.
复杂，复杂性；错综复杂的事物

**autonomous**/ɔːˈtɒnəməs/ adj.
自治的；自主的；自发的

**communication**
/kəmjuːnɪˈkeɪʃ(ə)n/ n.
通信；交流；信函

area for all kinds of AI core technologies: Mobile, autonomous agents operating in uncontrolled environments, stand-alone or in *collaborative* settings, allow to *investigate*, test and exploit technologies from robotics, computer vision, sensing, and environment interaction. Integrating multiple partners and their *heterogeneous* information sources leads to application of semantic technologies. The complexity of the agricultural production asks for progress in modeling capabilities, handling of uncertainty, and in the *algorithmic* and *usability* aspects of location- and context-specific decision support. The growing interest in reliable predictions as a basis for planning and control of agricultural activities requires the *interdisciplinary* cooperation with domain experts e.g. from agricultural research. Modern agricultural machines shall use self-configuring components and shall be able to collaborate and exhibit aspects of self-organization and swarm intelligence.

　　After an intensive reviewing, we selected four technical *contributions*, three reports, and two doctoral theses summaries, which are complemented by an interview with Joachim Keizer from FAO on THE ROLE OF LINKED DATA IN AGRICULTURE.

　　All technical papers in this Special Issue give a first view on this challenging interplay between AI and Agriculture. Taking profit from state-of-the-art sensing and actuator technologies the contribution on DATA MINING AND PATTERN RECOGNITION IN AGRICULTURE addresses challenges and potentials of appropriate methods in Agriculture. Motivated by the need for increased *resource* efficiency, the paper on ROBOTS FOR FIELD OPERATION WITH COMPREHENSIVE MULTILAYER CONTROL *summarizes* work on the development of autonomous agricultural machines. A contribution to better understanding between multiple cooperating actors is proposed in a submission on ONTOLOGY-BASED MOBILE COMMUNICATION. Optimizing the operation of a harvesting logistics chain, consisting of multiple cooperating vehicles in the field, will profit from the *application* of dynamic route planning algorithms, as presented in a paper on SPATIAL-TEMPORAL CONSTAINT PLANNING.

　　While the report on the iGREEN project spans from support for sharing and exchange among agricultural *operators* to decision support and application control, the re- port on TOWARDS SUPPORTING MOBILE BUSINESS PROCESSES focuses on the

*New Words and Expressions*

**collaborative**/kəˈlæbəreɪtɪv/ adj.
合作的，协作的

**investigate**/ɪnˈvestɪgeɪt/ v.
调查；研究

**heterogeneous**
/ˌhet(ə)rə(ʊ)ˈdʒiːnɪəs/ adj.
多相的；异种的；不均匀的；由不同成分形成的

**algorithmic**/ˌælgəˈrɪðmɪk/ adj.
算法的；规则系统的

**usability**/ˌjuːzəˈbɪləti/ n.
合用，可用；可用性

**interdisciplinary**
/ˌɪntəˈdɪsɪplɪn(ə)rɪ/ adj.
各学科间的；跨学科的

**contribution**/kɒntrɪˈbjuːʃ(ə)n/ n.
贡献；捐献；投稿

**resource**/rɪˈsɔːs/ n.
资源，财力；办法；智谋

**summarize**/ˈsʌməraɪz/ v.
总结；概述

**application**/ˌæplɪˈkeɪʃ(ə)n/ n.
应用；申请；应用程序；敷用

**operators**/ˈɒpəreɪtəz/ n.
经营者；操作者；[计] 运算符

uncertainty encountered in the non-deterministic agricultural *environment* and the application of agent technology to cope with that. Innovative ways for agricultural agents to see and perceive their environment are described in DETECTION OF FIELD STRUCTURES, which combines laser scanners and computer vision with sophisticated modeling capabilities to enable the intended structure recognition. In addition, progress results in successful and interesting doctoral dissertation work: In order to enable self-organized sensor integration in modular machines, BIO-INSPIRED SENSOR DATA MANAGEMENT took *inspiration* from ant colonies and similar observations. MECHATRONIC SYSTEMS investigates the adaptation of the operating parameters of a modern agricultural machine to the current context and task details in the field.

The overview given in this special issue is far from complete. On one hand, we only present work in the context of agricultural plant production. Cattle breeding, e.g., offer interesting examples for applied robotics or advanced sensing and crowd monitoring—none of this is covered in this issue. On the other hand, the articles are very much focused on individual actors or machinery and their interaction and collaboration within the initial production. Multistage processes, involving finishing and sales organizations (think about mills, bakery, and retail shops) or sustained operation over the year (see e.g. the operation of a biogas plant) are not covered. There is also no contribution on the handling of extra-agricultural issues like e.g. the *financial* markets.

*Nevertheless*, the collection of articles shows that Agriculture is a promising application field for AI technology, and in turn AI has a rich variety of important contributions to offer to cope with the pressing challenges faced by Agriculture. We are looking forward to further interesting work to our mutual benefit!

**New Words and Expressions**

**environment**/ɪnˈvaɪrənmənt/n.

环境，外界

**inspiration**/ˌɪnspəˈreɪʃn/ n.

灵感；鼓舞；吸气；妙计

**financial**/faɪˈnænʃ(ə)l/ adj.

金融的；财政的，财务的

**nevertheless**/nevəðəˈles/ adv.

然而，不过；虽然如此

## 参考译文 B

农业和人工智能这两个领域似乎是割裂的两个世界。农业是对土壤进行耕作、加工和生产粮食，特别务实。相比之下，人工智能与计算机系统、复杂的交互、建模和推理方法深深交织在一起，在公众的认知中仍是科幻小说风格的一部分。那么，我们为什么要把二者放在一起呢？

现代农业面临着巨大的挑战。今天，农业已发展成为一个高度竞争和全球化的工业，农民和其他行动者必须考虑当地气候和地理方面以及全球生态和政治因素，以保证经济生存和可持续生产。养活不断增长的世界人口需要不断增加粮食产量，但耕地仍然是一种有限的资源。对生物能源或改变饮食偏好的新要求给农业生产带来了额外的压力，而房屋和运输消耗了越来越多的土地。全球气候的预期、降雨模式的变化、全球变暖、干旱或极端天气事件发生的频率和持续时间的增加，对传统产区构成了威胁，并给全球粮食产量带来新的风险和不确定性。应对这些挑战，农业需要可持续性，在农业生产上要提高生产力和效率，而像水、能源、化肥等资源需要谨慎使用，有效地保护环境维持耕地土壤的质量与密度。其他难以预测的短期事件，如流行病、金融危机或农业原材料和产品的价格波动，加大了挑战的复杂性。

因此，农业生产需要处理复杂性、不确定性和模糊性，它要求农业生产的各个方面都有新的解决方案——从更好和可预测的作物规划，到精确耕作，优化资源应用，利用现代技术支持高效和协作过程，完全或部分自主解决烦琐工作，能持续长期发展有用的知识资源。

在近几十年里，人们实现了用计算机解决农业中这些挑战的基础：全球定位系统（GPS）（提供精确的位置数据，并为各种特定位置的支持提供基础）和移动通信（即使在现场，参与者之间也可以快速交换数据）是关键和公认的突破性技术。从现在可用的数据中获得意义，并使用产生的知识在各个层次上改进过程和操作，使 AI 及其建模和推理能力发挥作用。

从人工智能的角度来看，农业为各种人工智能核心技术提供了广阔的应用领域：移动的、在不受控制的环境中运行的、独立的或协作的设置，允许调查、测试和利用来自机器人、计算机视觉、传感和环境交互的技术。将多个合作伙伴及其异构信息源集成到一起，就可以应用语义技术。农业生产的复杂性要求在算法和可用性方面有所进展，包括建模能力，处理不确定性以及在特定的位置和上下的决策支持。人们对可靠预测作为规划和控制农业活动的基础越来越感兴趣，这要求各领域专家（如农业研究专家）进行跨学科合作。现代农业机械应使用自配置的部件，并能够协作和展示自组织和群体智能的各个方面。

经过深入审查，我们选择了 4 份技术介绍、3 份报告和 2 份博士论文摘要，并与粮农组织的约阿希姆·凯泽就相关数据在农业中的作用进行了访谈。

本期特刊的所有技术论文都对人工智能与农业之间的这种具有挑战性的相互作用给出了初步的看法。得益于最先进的传感技术和执行器技术，《农业数据挖掘和模式识别》一文介绍了应用于农业中的方法所面临的挑战和未来的潜力。在需要提高资源效率的激励下，论文《多层控制的野外作业机器人》综述了自主农业机械的研究进展。在《基于本体的移动通信》中，提出了一种有助于更好地理解多个合作参与者的方法，对一个由多个合作节点组成的收获物流链进行优化操作，将受益于动态路径规划算法的应用。

虽然关于 iGREEN 项目的报告涵盖了从支持农业操作系统之间的共享和交换，到决策支持和应用控制，《面向移动业务流程的支持》着重讨论了在不确定性农业环境中遇到的不确定性，以及如何应用智能体技术来应对这种不确定性。《土壤结构检查》描述了观察和感知农业环境的创新方法，该技术结合激光扫描仪和计算机视觉与复杂的建模能力，以实现预期的结构识别。此外，还有很多博士论文关注这一进展。为了能够在模块化机器中实现自组织传感器集成，《仿生传感器数据管理》一文从蚁群和类似的观察中获得了灵感。《机电一体化系统》研究了现代农业机械的操作参数对当前环境和任务细节的适应性。这期专刊所作的概述远不能令人满意。一方面，我们只在农业植物生产的背景下提出工作。例如，应用机器人或先进的传感器和人群监测等有趣的例子，没有一个能被这个问题涵盖。另一方面，这些文章在很大程度上是关于个体演员或机器以及他们在最初生产中的相互作用和协作的。多阶段的过程，包括精加工和销售组织（磨坊、面包房和零售商店）或在持续的操作（如沼气工厂的操作）都不包括在内。在处理农业以外的问题，如金融市场方面也没有贡献。

然而，本期特刊表明，农业是人工智能技术的一个很有前途的应用领域，而人工智能应对农业生产所面临的巨大的挑战但会做出重要的贡献。

# Chapter *7*

# Artificial Intelligence in Education

## Text A

Artificial Intelligence has the potential to greatly improve and change education systems across the world. There is a strong possibility for artificial intelligence to be used to help teachers *effectively* streamline their instruction process and to help students receive much more personalized help that is specifically suited to their strengths and *weaknesses*. Ideally AI will also help to complete some of the more menial tasks that teachers and teaching assistants have to work on, freeing them up to spend even more time helping students. There has been a lot of progress in this field in the past few years, as more and more companies have been involved in projects that aim to augment, improve, and change the way teaching is done. The field of Education is definitely ripe for innovation, and the *advancement* of artificial intelligence may be able to provide that *innovation*. The right applications of AI can result in students having a vastly more detailed and substantial education, as AI program can do way more to *identify* and target their individual strengths and weaknesses, and teachers can be aided by AI in order to have a larger amount of impactful teaching time with their students. Overall, education systems all over the world stand to greatly benefit from the proper *integration* of artificial intelligence in their schools.

| New Words and Expressions |
| --- |
| **artificial**/ɑːtɪˈfɪʃ(ə)l/ adj. |
| 人造的；仿造的；虚伪的；非原产地的；武断的 |
| **intelligence** /ɪnˈtelɪdʒ(ə)ns/ n. |
| 智力；情报工作；情报机关；理解力 |
| **effectively** /ɪˈfektɪvlɪ/ adv. |
| 有效地，生效地；有力地；实际上 |
| **weakness**/ˈwiːknəs/ n. |
| 弱点；软弱；嗜好 |
| **advancement**/ədˈvɑːnsm(ə)nt/ n. |
| 前进，进步；提升 |
| **innovation**/ˌɪnəˈveɪʃn/ n. |
| 创新，革新；新方法 |
| **identify**/aɪˈdentɪfaɪ/ v. |
| 确定；鉴定；识别，辨认出；使参与；把…看成一样 |
| **integration**/ˌɪntɪˈɡreɪʃ(ə)n/ n. |
| 集成；综合 |

Figure 7-1

**New Words and Expressions**

**personalize**/ˈpɜːs(ə)n(ə)laɪz/ v.

　使个性化；把…拟人化

**pioneering**/paɪəˈnɪərɪŋ/ adj.

　首创的；先驱的

**struggle**/ˈstrʌg(ə)l/ v.

　奋斗，努力；挣扎

**feasible**/ˈfiːzɪb(ə)l/ adj.

　可行的；可能的；可实行的

**menial**/ˈmiːnɪəl/ adj.

　卑微的；仆人的；适合仆人做的

**currently**/ˈkʌrəntlɪ/ adv.

　当前；一般地

**chunk**/tʃʌŋk/ n.

　大块；矮胖的人或物

**assignment**/əˈsaɪnm(ə)nt/ n.

　分配；任务；作业；功课

**implementation**

/ˌɪmplɪmenˈteɪʃ(ə)n/ n.

　实现；履行；安装启用

### Personalized Education with AI

AI can provide a great deal of benefits in the way that it has the potential to help schools *personalize* education for their students. There are many companies currently exploring the possibilities of AI being used to help tutor students. Companies such as Carnegie Learning , Thinkster[1] Math and Socratic are *pioneering* in this field, as they have developed services designed to help tutor students with their homework as well as the areas of math that they *struggle* with. The benefit to this type of service is that it allows students to a get something closer to a 1 on 1 teaching experience that is able to identify their strengths and weaknesses in a low pressure and convenient environment. It is simply not *feasible* for a teacher to be able to provide this level of deep personalized teaching to each student, but with more advanced AI programs, schools will be able to provide their students with that type of personalized teaching from AI and then their teachers will be able to step in to provide more help once the AI has done the basic teaching and identified which concepts they struggle with and which ones they understand just fine.

### AI Could Be Used to Grade Students' Work

AI can also do a great deal to help the education system by aiding teachers with some of their more *menial* tasks. *Currently*, teachers have to take a pretty solid *chunk* of time out of their days in order to grade *assignments*, tests and papers, as well as do other sorts of basic bookkeeping. With the proper *implementation*, an AI system could eventually be able to step in and take care of this sort of work. Automating a solid amount of a teacher's more menial tasks would open up a substantial portion of their schedule, allowing them to focus on more significant quality time with students and give them more time to focus on their lesson plans and on the statuses of each of their students. This would all be great for the students' education because they would have more quality time to spend being *personally* instructed by their teacher.

Figure 7-2

Ultimately, there is a fantastic amount of good things that AI can do for education systems if properly implemented. Numerous companies are working on exciting AI programs that are positioned to add a great deal to the educational experience for both teachers and their students. AI programs can be integrated into schools in order to help students study their coursework at all times, as well as identify which *mathematical* concepts they are struggling with and provide them with extra support in order to better understand the points that they are struggling with. There are also exciting possibilities in terms of AI programs being able to *automate* the more menial tasks that teachers currently handle *manually*. The automation of these tasks would open up a great deal of time for teachers to spend on more valuable parts of the educational experience. Overall, artificial intelligence can greatly improve education systems through its ability to streamline many parts of a teacher's job, and automate other parts, ultimately giving them more and more time to spend directly on their students.

**New Words and Expressions**

**personally**/ˈpɜːs(ə)n(ə)lɪ/ adv.
　亲自地；当面；个别地；就自己而言

**mathematical**
/mæθ(ə)ˈmætɪk(ə)l/ adj.
　数学的，数学上的；精确的

**automation**/ɔːtəˈmeɪʃ(ə)n/ n.
　自动化；自动操作

**manually**/ˈmænjʊəli/ adv.
　手动地；用手

## Terms

智囊团（Thinkster）又称头脑企业、智囊集团或思想库、智囊机构、顾问班子。Thinkster是指专门从事开发性研究的咨询研究机构。它将各学科的专家学者聚集起来，运用他们的智慧和才能，为社会经济等领域的发展提供满意方案或优化方案，是现代领导管理体制中的一个不可缺少的重要组成部分。

## Comprehension

### Blank Filling

1. _____ can improve and change teaching methods.

2. Artficial intelligence can help schools provide _____ education for students.

3. _____ can be used to grade students' homework.

4. Many companies are working on _____.

5. _____ can automatically complete more trivial tasks that teachers handle manually.

6. _____ can simplify teachers' work.

### Content Questions

1. What is the definition of Artificial intelligence?

2. What are the characteristics of Artificial intelligence?

3. What is the background of Artificial intelligence?

## Answers

### Blank Filling

1. Artificial intelligence

2. Personalized

3. Artificial intelligence

4. Artificial intelligence projects

5. Artificial intelligence programs

6. Artificial intelligence

### Content Questions

1. "Artificial intelligence" It is a new technology science to research and develop theories, methods, technologies and application systems for simulating, extending and extending human intelligence.

2. There is no professional limitation: artificial intelligence belongs to the field of computer science research, but AI is the study of machine intelligence. Wherever the human brain is used, AI can be used. Therefore, AI can be applied in various professional fields.

3. Artificial intelligence is a very challenging science, people who do this work must know computer knowledge, psychology and philosophy. Artificial intelligence includes a very wide range of science, it consists of different fields, such as machine learning, computer vision and so on, in general, one of the main goals of artificial intelligence research is to make machines capable of some complex tasks that usually require human intelligence to complete.

## 参考译文 A

人工智能极大地改善和改变世界各地的教育系统。人工智能很有可能被用来帮助教师有效地简化他们的教学过程，并帮助学生得到更个性化的帮助。理想情况下，人工智能还能帮助教师和助教完成一些更琐碎的工作，让他们腾出更多时间帮助学生。在过去的几年里，这一领域已经取得了很大的进步，越来越多的公司参与了改进和改变教学方式的项目。教育领域的创新显然已经成熟，而人工智能的进步或许能够提供这种创新。正确的人工智能应用程序可能会为学生提供一个更详细和实质性的教育，人工智能程序能够识别目标的优势和劣势，教师可以通过人工智能的帮助，与他们的学生有更多的有效教学时间。总的来说，全世界的教育系统都将从人工智能中获益良多。

### 人工智能个性化教育

人工智能益处颇多，它有潜力帮助学校为学生提供个性化的教育。目前有许多公司正在探索人工智能辅导学生的可能性。卡耐基学习（Carnegie Learning）、智库（Thinkster）和家庭作业助手（Socratic Math）等公司在这一领域处于领先地位，因为它们开发出了旨在帮助指导学生完成作业以及解决它们所遇到的数学问题的服务。这类服务的好处在于，它能让学生在低压力和便利的环境中找到自己的优势和劣势。对于老师来说，为每个学生提供深度个性化教学是不可行的。但在更高级的人工智能程序中，学校能够为学生提供人工智能类型的个性化教学，他们的老师也能够介入并提供辅助帮助。

### 人工智能可以用来给学生的作业评分

人工智能还可以通过帮助教师完成一些更琐碎的任务，为教育系统提供很大帮助。目前，教师们不得不从日常生活中抽出相当多的时间来批改作业、考试和论文，以及做其他一些基本的簿记工作。人工智能系统最终能够介入并处理这类工作，使他们能够专注于与学生相处，并更高效地将教师大量的较琐碎的工作自动化，人工智能系统将在他们的课程表中占据相当大的一部分，并让教师有更多的时间专注于他们的课程计划和每个学生的状态。这对学生的教育来说是很好的，因为他们有更多的时间是由老师亲自指导的。

最终，如果得到合理的实施，人工智能可以为教育系统做很多好事。许多公司都在致力于令人兴奋的人工智能项目，这些项目的定位是为教师和学生提供更多的教育体验。人工智能项目可以整合到学校中，帮助学生在任何时候都可以学习他们的课程，并为他们提供额外的支持，以便更好地理解他们正在努力学习的要点。人工智能的这些应用确实很让人兴奋，它能够自动完成教师目前手工处理的更琐碎的任务。这些任务的自动化将为教师节省大量的时间，让他们花在更有价值的教育经历上。总的来说，人工智能可以极大地改善教育系统，因为它能够简化教师工作的许多部分，并使其他部分自动化，使教师越来越多的时间花在学生身上。

# Text B

The *application* of AI methods to problems such as legal decision making, language translation, or gene analysis often requires the *cooperation* of AI experts and subject specialists, e.g., lawyers, translators, or *biologists.* Their ability to communicate on a common ground is a crucial factor determining the success of the project. It is thus *beneficial* if both parties have a basic understanding of the subject as well as of AI methods, even before the start of a project.

Universities provide a unique *opportunity* to both teach students becoming AI experts some subject knowledge (e.g., biology or law) and ensure that students in non-computing subjects have a basic understanding of AI techniques. A native approach for achieving such *interdisciplinary* learning is that AI students take some first-year subject courses, and subject students some introductory AI courses. Even though this approach is easy to implement, it may not achieve the intended interdisciplinary learning benefits since the courses are not tailored towards students of a different *discipline* (even first-year courses often provide a detailed introduction to a specific topic instead of surveying a whole field).

We here discuss two approaches based on peer-learning, which provide a more beneficial interdisciplinary learning environment. They share the idea that AI and subject students learn together by teaching each other.

In the seminar-style approach, AI students give seminars to subject students (and vice versa). These seminars may, for example, provide an overview of AI techniques or review applications of AI methods in subject areas. This approach does not only benefit the attending subject students, who acquire knowledge tailored particularly to them, but also provides valuable experience to the AI student giving the seminar in explaining AI topics to the lay *audience.* There is clearly a lot of *variability* concerning the exact setup of these seminars: they can be given by a single PhD student or by a group of *undergraduates*, and the attendees' background can be a mixture of subjects or a single subject (in which case the seminar will cover topics and examples related to this particular subject).

In contrast to the seminar-style approach, where the speaker

---

**New Words and Expressions**

**application**/ˌæplɪˈkeɪʃ(ə)n/ n.
应用；申请；应用程序；敷用

**cooperation** /kəʊˌɒpəˈreɪʃ(ə)n/ n.
合作，协作；协力

**biologist**/baɪˈɒlədʒɪst/ n.
生物学家

**beneficial**/benɪˈfɪʃ(ə)l/ adj.
有益的，有利的；可享利益的

**opportunity**/ˌɒpəˈtjuːnətɪ/ n.
时机，机会

**interdisciplinary**
/ɪntəˈdɪsɪplɪn(ə)rɪ/ adj.
各学科间的；跨学科的

**discipline**/ˈdɪsɪplɪn/ n.
学科；纪律；训练；惩罚

**audience**/ˈɔːdɪəns/ n.
观众；听众；读者；接见；正式
会见；拜会

**variability**/ˌveərɪəˈbɪlətɪ/ n.
可变性，变化性；变异性

**undergraduate**/ʌndəˈɡrædjʊət/ n.
大学生；大学肄业生

teaches the audience, the project-based approach promises *mutual* teaching and learning, both in terms of knowledge and skills. In this setting, an AI student and a subject student work together on a project trying to solve a problem in the subject student's area by applying AI techniques. At the start, the subject student explains subject-specific *background* to the AI student, whereas the AI student teaches the subject student about possible AI techniques to be used, thus creating a mutual teaching and learning environment. During the project, students will also acquire the invaluable skills of working in an interdisciplinary team. Again, there are different setups for such projects: The problem(s) to be solved can be given by faculty or be the students' own ideas, and the project can be part of a course or an extracurricular "ideas/start-up lab".

Most approaches in university teaching are based on frontal *lectures*, sometimes with specific lab activities and specific homeworks. The course is divided in specific modules which are explained *sequentially*.

I would be interested in analysing the *feasibility* (and try that with a real course) of a more student-centric approach, inspired by the *pedagogical* Montessori method (Montessori and George 1964). Although the main focus of the method has always been on children, some of those elements have been incorporated with success in secondary school and early-undergraduate levels.

Working with an equipped lab is *fundamental* for this approach. Then I would imagine that each student (or maybe each group of students) could freely decide the direction of the course, based on discovery and on what they are interested in. I have recently started a cooperation with Prof. Federico Gobbo at the University of Amsterdam, to analyze the *portability* of some key elements of the Montessori method into AI education of young adults (so, the target group of this call).

From one side, I am interested to see how the Montessori method applied at a later age group than usual could help the students in their personal and professional development. I strongly believe that independence and the ability of thinking, reasoning, and making informed choices are key elements of the lives of active and engaged human beings, part of the society. A teaching approach which values independent thinking seems therefore a very interesting

**New Words and Expressions**

**mutual**/ˈmjuːtʃʊəl/ adj.
共同的；相互的，彼此的
**background**/ˈbækɡraʊnd/ n.
背景；隐蔽的位置
**lecture**/ˈlektʃə/ n.
演讲；讲稿；教训
**sequentially**/sɪˈkwɛnʃəli/ adv.
从而；继续地；循序地
**feasibility**/fiːzɪˈbɪlɪtɪ/ n.
可行性；可能性
**pedagogical**/ˌpedəˈɡɒdʒɪkl/ adj.
教育学的；教学法的
**fundamental**/fʌndəˈment(ə)l/ adj.
基本的，根本的
**portability**/ˌpɔːtəˈbɪlətɪ/ n.
可移植性；轻便；可携带性

and potentially fruitful approach, albeit maybe difficult at times.

From the other side, looking for new engaging methods of teaching AI and robotics might result in students approaching the subject with curiosity and willingness, not just because it is in the study plan. This in turn might result in more people engaged in AI and Robotics, and in more passion towards the subject. It might be perceived not just as one of many lectures, but a feel of "ownership" might push for a deeper understanding of specific subjects rather than usual frontal lectures.

Challenges in this approach would be ensuring that each student (or each group of students) progress and explore the subject within some boundaries. Also, evaluation is a very delicate subject. In the original Montessori approach there is no grading for children, but it is something usually necessary in undergraduate courses. Establishing a fair grading system is something necessary, though it might be hard to compare different approaches and different paths that each student would undertake.

AI can also teach us about what it means to be human. It can teach us what *humanity* looks like when taken to different extremes and thus develop within ourselves a deeper understanding of each other and our differences. It can easily demonstrate the truth and beauty of mathematics and how it can be used to develop models of knowledge and behavior. Each one of these models can then provide us with a unique *perspective* into our own cognition, *psychology* and the perspective of our *existence*.

A solid foundation in mathematics will start with movement, which will flow from real object *manipulation* to imagination to abstract cognition. This is introduced with early arithmetic. However, there is no similar early introduction of non-arithmetic *cognition* such as logic, search, iteration (folding), etc... that are vital for all kinds of engineering and programming. Such professions are shown to be deeply imaginative from mentally stepping through a program's *execution* to predicting the voltage levels across a circuit diagram. Early introduction of agent based models through games and puzzles could provide this foundation as well as begin to introduce concepts for later exploration such as search, string-replacement-iteration, planning, machine learning, etc.

What this solid foundation of movement and imagination

**New Words and Expressions**

**humanity**/hjʊˈmænɪtɪ/ n.

人类；人道；仁慈；人文学科

**perspective**/pəˈspektɪv/ n.

观点；远景；透视图

**psychology**/saɪˈkɒlədʒɪ/ n.

心理学；心理状态

**existence**/ɪgˈzɪst(ə)ns/n.

存在，实在；生存，生活；存在物

**manipulation**/məˌnɪpjʊˈleɪʃ(ə)n/ n.

操纵；操作；处理；篡改

**cognition**/kɒgˈnɪʃ(ə)n/ n.

认识；知识；认识能力

**execution**/ˌeksɪˈkjuːʃ(ə)n/ n.

执行，实行；完成；死刑

provides is a deeper understanding of and greater passion for mathematics. By the time we get to techniques such as multiple column multiplication or long division we are beginning to learn procedures. This is what will make or break the love of mathematics. Those that truly learn what is behind the procedure and can see it in their imagination will do well, those that learn to blindly follow the procedure will not.

I hope to explore the earliest introduction of core AI concepts in a concrete way to develop technical imagination skills, get us to think about how we think and finding new confidences in ourselves as we explore what it means to be human.

While it may be unreasonable to expect early undergraduate and secondary school students to code AI algorithms, it is possible for them to visualize and experience these algorithms firsthand. Developing an understanding of AI through these perspectives may even facilitate abstract thinking and problem solving when learning computer science and programming later. Although taught later in the CS curriculum after students are comfortable with computational thinking, many topics in AI can be explained conceptually using only high school mathematics. However, the manner in which these concepts are taught needs to be less traditional.

Based on the average student's present-day lifestyle involving personal mobile devices and almost limitless access to media, most students are used to constantly interacting with others and engaging in entertainment. This nearly contradicts the traditional lecture style for presenting material impersonally at the front of the room using chalkboards or slides. Instead, students today are accustomed to short spurts of watching and then lots of time doing, which goes hand-in-hand with some elements of team-based learning. In particular, an instructor should only briefly introduce a topic and related activity. Then, the students may explore the activity in groups in order to experience the concept on their own, interacting with each other to understand what happens. For example, a search can be performed with a map and deck of cards; each card covers a city and students write the ruler distance (Euclidean heuristic) on each card as it is added to the frontier. The visited cities' cards are stacked in a deck to visualize the visited sequence.

By focusing on the algorithms' processes rather than the specific implementation, younger students without computational experience, higher-level mathematics, and programming skills can participate. The early focus of AI was to emulate human intelligence, and these students can relate to that by wondering, "how would I solve this problem?" These are questions they can discuss with each other and the instructor while performing the activities. In particular, the instructor can now make her time with students more personal by visiting groups to discuss and give tips based on their progress. Groups can also interact with each other afterwards to compare results.

Just as important as the interaction in the classroom, time outside of class can be vital to learning. Besides homework assignments that review concepts, students spend time on the internet watching videos and listening to music. Educational content can be provided in such entertaining forms. Younger students are exposed to AI topics at any time in formats that they are more ready to digest, using high school-level knowledge without focusing on the code.

**Note:**

The text is adapted from the website:

https://hrilab.tufts.edu/publications/eaton2017EAAI.pdf

# 参考译文 B

　　人工智能方法在诸如合法决策、语言翻译或基因分析等问题上的应用常常需要人工智能专家和学科专家（如律师、翻译或生物学家）的合作。他们在共同基础上进行沟通的能力是决定该项目成功与否的一个关键因素。因此，在项目开始之前，双方对该主题以及 AI 方法有一个基本的了解是非常有益的。

　　大学为想成为人工智能专家的学生提供了一个独特的机会，教授他们一些学科知识（如生物或法律），又确保非计算机专业的学生对人工智能技术有一个基本的了解。实现这种跨学科学习的一个方法是采用一年级主题课程。尽管这种方法很容易实施，但它可能无法达到预期的跨学科学习效果，因为这些课程不针对不同学科的学生（即使是第一年的课程，也往往提供一个特定主题的详细介绍，而不是调查整个领域）。

　　我们在这里讨论两种基于同行学习的方法，它们提供了一个更有益的跨学科学习环境。他们分享了人工智能和学科学习通过互相教学来共同学习的理念。

　　在研讨会式的方法中，人工智能学生为主题学生提供研讨会（反之亦然）。例如，这些研讨会可以概述人工智能技术或审查人工智能方法在学科领域的应用。这种方法不仅对主课的学生有好处，因为这需要专门为他们量身定制的知识，而且也为人工智能学生提供了宝贵的经验，让他们能够通过讨论会向普通观众解释人工智能的主题。

　　与研讨会式的教授方法不同，基于项目的方法在知识和技能方面保证了相互教授和学习。在这种情况下，一个人工智能学生和一个学科学生一起在一个项目工作，试图通过应用人工智能技术解决学科学生领域的问题。首先，在这种具体背景，被试教授可能用到的人工智能技术，从而创造一个互教互学的环境。在项目中，学生们在跨学科团队中交流工作中的宝贵技能。同样，这类项目有不同的设置：要解决的问题可以由教师提出，也可以由学生自己提出，项目可以是课程或课外"创意/创业实验室"的一部分。

　　在大学教学中，大多数方法都是建立在正面讲课的基础上，有时还包括具体的实验活动和作业。本课程以特定的模式来划分，并依序加以解释。

　　我分析了一种以学生为中心的教授方法的可行性（并在一门真正的课程中尝试），这种方法的灵感来自蒙氏教学法(Montessori and George 1964)。虽然该方法的重点一直是儿童，但其中一些因素已经与中学和本科早期阶段成功地结合在一起。

　　与装备齐全的实验室合作是这项研究的基础。我可以想象每个学生（或者每一组学生）都可以根据自己的发现和兴趣自由决定课程的方向。我最近开始与阿姆斯特丹大学的 Prof. Federico Gobbo 合作，分析 Montessori 方法的一些关键要素在人工智能教育中具有可移植性。

　　一方面，我很感兴趣的是蒙氏索利法在一个较晚的年龄组的应用，探讨如何在他们的个人和专业发展中帮助学生。我坚信独立和思考、推理和做出明智选择的能力是人们生活的关键因素。因此，一种重视独立思考的教学方法似乎是一种非常有趣且富有成效的方法，尽管有时可能很困难。

另一方面，寻找教授人工智能和机器人技术的新方法，可能会引导学生有好奇心和意愿来学习这门学科，而不仅仅是因为它在学习计划中。这反过来可能会使更多的人从事人工智能和机器人技术，并对这一学科产生更多的热情。它可能不仅被视为众多讲座中的一个，而且一种"所有权"的感觉可能会推动人们对特定主题的更深入理解，而不是通常的直接讲授。

这种方法的挑战将是确保每个学生（或每一组学生）在一定范围内进步和探索该主题。此外，评价是一个非常微妙的问题。在最初的蒙特梭利教学方法中，没有针对儿童的评分，但在本科课程中通常是必要的。建立一个公平的评分系统是必要的。

人工智能还可以教会我们做人的意义。它可以教会我们，当我们面对不同的极端情况时，人性是什么样子的，从而使我们对彼此之间的差异有了更深的了解。它可以很容易地展示数学的真和美，以及如何用它来建立知识和行为的模型。这些模型中的每一个都可以为我们提供一个独特的视角来了解我们自己的认知和心理。

一个扎实的数学基础要从运动开始，运动要从对真实物体的操纵，到想象，再到抽象的认识。在非算术认知领域，如逻辑、搜索、迭代（折叠）等，早期是没有类似的内容的。这对所有类型的工程和编程都是至关重要的。这类知识显示出了深刻的想象力，从心理上一步步地执行一个程序，到通过电路图预测电压水平。早期通过游戏和谜题引入类似的模型可以为以后的探索打下基础，并开始引入概念，如搜索、字符串替换迭代、计划、机器学习等。

这种坚实基础提供了学习数学的热情。当我们学习多列乘法或长除法等知识时，我们已经开始学习过程了。我希望尽早以具体的方式介绍人工智能核心概念，以培养我们的技术想象力，使我们如何思考，并在探索做人的意义时找到新的自信。

我希望以一种具体的方式探索引入核心人工智能概念的方法，以发展技术想象技能，让我们思考并探索人类。

虽然期望早期的大学生和中学生编写人工智能算法是不合理的，但他们有可能将这些算法可视化并亲身体验。通过这些视角对人工智能进行理解，甚至可以帮助学生在学习计算机科学和编写程序时进行抽象思考和解决问题。虽然在计算机科学课程中，在学生适应计算思维之后才会学习，但是 AI 中的许多主题只能用高中数学来进行概念上的解释。然而，教授这些概念的方式往往也不那么传统。

根据当今普通学生的生活方式——移动个人设备和几乎无限的媒体接入，大多数学生习惯于不断地与他人互动和进行娱乐活动。这几乎与传统的用黑板或幻灯片在教室前面客观地呈现讲课方式相矛盾。相反，今天的学生习惯了短时间的观看，然后用大量的时间做，这与团队学习的一些元素是密切相关的。特别是，教师只需要简要介绍一个主题和相关活动。学生可以在小组中探索交流，以亲身体验概念，了解发生了什么事。例如，可以使用地图和卡片进行搜索；每张卡片覆盖了一个城市，学生们在每一张卡片上写下"标尺距离"（欧几里得启发式）。被访问城市的卡片被堆叠在一副牌上，以可视化被访问城市的顺序。

通过关注算法过程，而不是具体的实施，让没有电脑操作经验、没有高等数学和程序设计技能的年轻学生也可以参加。人工智能早期的重点是模仿人类的智力，这些学生可以通过思考"我如何解决这个问题""这些问题是他们在活动中可以互相讨论或和指导老师讨论的"，老师可以与他的学生访问小组进行讨论，并根据他们的进展给出提示。小组之间还可以相互交流，比较结果。

就像课堂上的互动一样，课外时间对学习也是至关重要的。除了复习家庭作业，学生们还花时间上网看视频和听音乐。教育内容可以以这种娱乐形式提供。年轻的学生在任何时候都可以接触到更容易消化的人工智能主题，而不只是专注于代码。

# Chapter *8*

## Artificial Intelligence in Security

## Text A

If you've been keeping an eye on the latest in tech news, you know that both artificial intelligence (AI) and *cybersecurity* were subjects of hot discussion in 2017. They're only going to get hotter in 2018.

What you might not know is that the development of both AI and cybersecurity could mean an end to cybersecurity as we know it. New threats and hacking techniques are eclipsing traditional cybersecurity approaches every day, while new machine learning and artificial intelligence techniques allow IT personnel to *combat* threats in real time.

IT pros are already using AI to ward off cyberattacks. In fact, using artificial intelligence to detect threats is how most *businesses* have incorporated AI into their IT *strategies*.

If you want to stay on the bleeding edge of AI tech to prepare for prescient threats, read on to learn how advanced cyberthreats are changing security needs and how your business can *leverage* AI to protect your data and your *profits*.

### Uncharted *Territory*

High-profile cyberattacks like the Equifax breach, voter records hack, and the Department of Homeland Security breach are *evidence* of the changing landscape of cybersecurity.

Cybersecurity has long hinged on a "sealed borders" approach, where threats are blocked using *firewalls*, *antivirus* software, and password protection. These techniques have proved inadequate over

> **New Words and Expressions**
> **cybersecurity**/ˈsaɪbəsɪˈkjʊərətɪ/ n.
> 网络安全
> **combat**/ˈkɒmbæt/ v.
> 战斗；搏斗
> **business**/ˈbɪznɪs/ n.
> 商业；【贸易】生意；交易
> **strategy**/ˈstrætədʒɪ/ n.
> 战略，策略
> **leverage**/ˈliːv(ə)rɪdʒ/ v.
> 利用；举债经营
> **profit**/ˈprɒfɪt/ n.
> 利润；利益
> **territory**/ˈterɪt(ə)rɪ/ n.
> 领土，领域；范围；地域；版图
> **evidence**/ˈevɪd(ə)ns/ n.
> 证据，证明；迹象；明显
> **firewall**/ˈfaɪəwɔl/ n.
> 防火墙
> **antivirus**/ˈæntɪvaɪrəs/ n.
> 反病毒程序；抗病毒素

Figure 8-1

**New Words and Expressions**

**investment**/ɪnˈves(t)m(ə)nt/ n.

  投资；投入；封锁

**individual**/ˌɪndɪˈvɪdʒʊəl/　n.

  个人，个体

**approximately**/əˈprɒksɪmətlɪ / adj.

  大约，近似地；近于

**employee**/ɪmˈplɔɪi/ n.

  雇员；从业员工

**weapon**/ˈwep(ə)n/ n.

  武器，兵器

the past decade, and breaches are increasing in frequency in every sector.

## Cybercrime By the Numbers

According to Cybersecurity Ventures CEO Steve Morgan, experts anticipate a significant increase in cybersecurity threats over the coming decade. Potential losses from cybercrime could exceed six trillion by 2021, despite an expected one trillion in cybersecurity *investment*. Data breaches were up 13% in the first half of 2017 alone, and the problem is slated to get worse.

In an interview with Gizmodo, Marc Goodman—cybercrime expert and author of Future Crimes—reveals that most cybersecurity attacks today are automated, from AI-powered Distributed Denial of Service (DDOS) attacks to phishing scams and ransomware. Techniques like automated spear phishing and whaling use social engineering to target *individuals* with access and power in a company, making potential breaches all the more damaging.

Morgan notes that the "attack surface" available to hackers will increase to *approximately* six billion people by 2022. This makes building a cybersecurity defense much more difficult as any *employee* or web-connected device represents a vulnerability.

## Artificial Intelligence in Cybersecurity

"Hackers have been using artificial intelligence as a *weapon* for quite some time," says Brian Wallace—lead security data scientist at cybersecurity firm Cylance—told Gizmodo.

AI solves a hacker's problem with scale. It can help hackers use social engineering techniques to implement a spear-phishing campaign, and identifies targets faster and more efficiently than a human *operator*.

*New Words and Expressions*

**operator**/ˈɒpəreɪtə/ n.

经营者；操作员；话务员；行家

**preventative**/prɪˈventətɪv/ adj.

预防性的；防止的

**eliminate**/ɪˈlɪmɪneɪt/ v.

消除；排除

**digital**/ˈdɪdʒɪt(ə)l/ adj.

数字的；手指的

**infiltrator**/ɪnˈfɪltreɪtə/ n.

渗入者，渗透者

**revolutionize**/ˈrɛvəˈluʃəˈnaɪz/ v.

发动革命；彻底变革；宣传革命

**anomalous**/əˈnɒm(ə)ləs/ adj.

异常的；不规则的；不恰当的

Figure 8-2

## AI-Powered Information Security

With an automated approach, it's nearly impossible for any *preventative* security measures to keep pace. But, this doesn't mean businesses are defenseless. The latest breakthroughs in information security can help re-secure the digital world.

Several startups in the cybersecurity sector are fighting fire with fire, leveraging artificial intelligence to detect and *eliminate* threats in real time, like white blood cells in an immune system.

Most of traditional cybersecurity focuses on preventing attacks through known methods and actors. AI-powered network defenses are the only answer for "unknown unknowns," says Nicole Eagan, CEO and founder of cybersecurity firm Darktrace. AI can learn how to spot anomalies and patterns in normal network activity, identifying a potential *infiltrator* when it encounters a previously unknown method or technique.

This active approach to cybersecurity is sparking an arms race in AI development as hackers and data security scientists try to out-develop each other to maximize profit. Artificial intelligence allows cybersecurity to expand in scale, speed, and accuracy, creating an effective defense against automated or AI-driven cyberattacks.

Machine learning algorithms can *revolutionize* cybersecurity networks and approaches in the following ways:

Detection: AI lets businesses track *anomalous* behavior within a network in real time. With the AI system processing all normal network behavior, human analysts are free to identify and track anomalous data.

Protection: AI can identify and prioritize vulnerabilities for preemptive protections and, in some cases, even address vulnerabilities directly. Through smarter preemptive action, costs can be streamlined and low-level threats such as phishing scams can be eliminated.

Prediction: AI can anticipate and plan for unknown techniques and strategies before they occur, keeping your firm one step ahead of potential hackers, and the rest of the industry.

Termination: AI can stop cyberattacks before your network is irreversibly damaged. As AI technology continues to advance, algorithms will "learn" from each attack, developing more sophisticated anti-malware techniques.

**The Future of Your Cybersecurity**

AI algorithms aren't self-sufficient, or self-aware. A trained team of human analysts is essential for maximizing AI's benefits.

Artificial intelligence can overcome many of the common problems associated with cybersecurity—through streamlining and automation, AI tools can help analysts use resources more effectively and efficiently.

**What to Ask When Looking for a Cybersecurity Vendor**

When vetting potential cybersecurity vendors for your firm, it's good to be skeptical of any claims about total AI integration. The technology simply isn't there yet; human analysts remain an essential part of the process.

Ask questions about the implementation of AI tools to address industry-specific problems. These questions will reveal your vendor's AI experience level.

Machine learning algorithms should be useful to a variety of professionals in your security operations center—from experienced analysts down through new staff—so ensure that your vendor can address expertise and knowledge gaps at every level.

AI is changing the landscape of cybersecurity threats. Stay informed, and be prepared to use it for protection against the very threats it poses.

## Comprehension

### Blank Filling

1. New threats and _____ are eclipsing traditional cybersecurity approaches every day, while new _____ and _____ techniques allow IT personnel to combat threats in real time.

2. Ai can help hackers use _____ to spearhead phishing.

3. The latest breakthroughs in information security are helping to restore security to the _____.

4. _____ can handle all normal network behavior.

5. Ai can help analysts use resources more efficiently by _____ and _____.

6. To gain the competitive advantage that big data holds, you need to infuse _____ everywhere, make speed a differentiator, and exploit _____ in all types of data.

**Content Questions**

1. What is "sealed borders" approach?

2. How AI is used in cybersecurity?

3. Say the differences between traditional and AI cybersecurity.

# Answers

**Blank Filling**

1. hacking techniques; machine learning; artificial intelligence

2. social engineering techniques

3. digital world

4. Ai systems

5. streamlining; automating

6. analytics; value

**Content Questions**

1. Cybersecurity has long hinged on a "sealed borders" approach, where threats are blocked using firewalls, antivirus software, and password protection. These techniques have proved inadequate over the past decade, and breaches are increasing in frequency in every sector.

2. AI solves a hacker's problem with scale. It can help hackers use social engineering techniques to implement a spear-phishing campaign, and identifies targets faster and more efficiently than a human operator.

3. Most of traditional cybersecurity focuses on preventing attacks through known methods and actors. AI-powered network defenses are the only answer for "unknown unknowns". AI can learn how to spot anomalies and patterns in normal network activity, identifying a potential infiltrator when it encounters a previously unknown method or technique.

# 参考译文 A

如果你一直关注着最新的科技新闻，你就会知道人工智能（AI）和网络安全在 2017 年都是热门话题。它们只会在 2018 年变得更热门。

你可能不知道的是，人工智能和网络安全的发展可能意味着我们所知道的网络安全的终结。新的威胁和黑客技术每天都在超越传统的网络安全手段，而新的机器学习和人工智能技术使 IT 人员能够实时应对威胁。

IT 专业人士已经在利用人工智能抵御网络攻击。事实上，利用人工智能来检测威胁是大多数企业将人工智能纳入 IT 战略的方式。

如果你想站在人工智能技术的最前沿，为可能的威胁做好准备，就必须首先了解先进的网络威胁如何改变安全需求，以及你的企业如何利用人工智能保护你的数据和利润。

### 未知的领域

像 Equifax 黑客事件、选民记录遭黑客攻击、国土安全部（Department of Homeland Security）遭黑客攻击等备受瞩目的网络攻击，都是网络安全形势不断变化的例子。

网络安全长期以来依赖于一种"封闭边界"的方法，即使用防火墙、杀毒软件和密码保护来阻止威胁。事实证明，这些技术在过去 10 年里并不完善，而且每个行业的违规次数都在增加。

### 网络犯罪的数字

据网络安全风险投资公司首席执行官史蒂夫·摩根称，专家预计未来十年网络安全威胁将大幅增加。到 2021 年，尽管用在网络安全上的投资预计将达到 1 万亿美元，但网络犯罪的损失有可能超过 6 万亿美元。仅在 2017 年上半年，数据泄露事件就增加了 13%，而且这一问题还将进一步恶化。

在接受 Gizmodo 采访时，网络犯罪专家、《未来犯罪》一书的作者 Marc Goodman 透露，如今大多数网络攻击都是自动化的，从人工智能的分布式拒绝服务（DDOS）攻击到钓鱼诈骗和勒索软件。像自动鱼叉式钓鱼和捕鲸这样的技术，利用社会工程来锁定公司中有权有势的个人，使得潜在的漏洞更具破坏性。

摩根指出，到 2022 年，黑客可利用的"攻击面"将增加到大约 60 亿人。这使得建立网络安全防御变得更加困难，因为任何员工或联网设备都代表着一个漏洞。

### 网络安全中的人工智能

网络安全公司 Cylance 的首席安全数据科学家 Brian wallac 告诉 Gizmodo："黑客使用人工智能作为武器已经有一段时间了。"

人工智能黑客借助人工智能扩大了规模。它可以帮助黑客使用社会工程技术来实施鱼叉式网络钓鱼活动，并比人类操作员更快、更有效地识别目标。

### AI-powered 信息安全

使用自动化方法，几乎不可能使任何预防性安全措施跟上进度。但这并不意味着企业毫无防备。信息安全方面的最新突破有助于恢复数字世界的安全。

网络安全领域的几家初创公司正在使用相同的方法，利用人工智能实时检测和消除威胁，就像免疫系统中的白细胞。

传统的网络安全主要是通过已知的方法和参与者来防止攻击。网络安全公司 Darktrace 的首席执行官和创始人妮可·伊根（Nicole Eagan）说，人工智能网络防御是"未知"的唯一答案。人工智能可以学习如何在正常的网络活动中发现异常，在遇到未知的方法或技术时识别潜在的问题。

随着黑客和数据安全科学家试图超越彼此，以实现利润最大化，这种积极的网络安全方法正在人工智能领域引发一场军备竞赛。人工智能使网络安全在规模、速度和准确性方面得以扩大，为抵御自动化或人工智能驱动的网络攻击创造了有效的防御手段。

### 机器学习算法可以通过以下方式革新网络安全

检测：AI 可以让商家实时跟踪网络中的异常行为。人工智能系统处理所有正常的网络行为，分析、识别和跟踪异常数据。

保护：人工智能可以识别并优先考虑漏洞，以便进行先发制人的保护，在某些情况下，甚至可以直接解决漏洞。通过更聪明的先发制人的行动，使成本可以被精减，如钓鱼诈骗可以被消除。

预测：在未知技术和策略出现之前，人工智能可以预测和计划它们，让你的公司领先于其他人一步。

终止：人工智能可以阻止网络攻击。随着人工智能技术的不断进步，算法将从每次攻击中进行"学习"迭代，开发出更复杂的反恶意软件技术。

### 你的网络安全的未来

人工智能算法不能自给自足，也不能自我感知。一个训练有素的人工分析师团队在如何让人工智能发挥最大化的作用方面至关重要。

人工智能可以克服许多与网络安全相关的常见问题，通过精简和自动化，人工智能工具可以帮助分析师更有效地利用资源。

### 在寻找网络安全供应商时应该问些什么

在为客户审查潜在的网络安全缺陷时，最好对任何关于人工智能给出的结果持怀疑态度。人工分析师仍然是这一过程的重要组成部分。

询问 AI 工具的实现，以解决特定行业的问题。这些问题将揭示您的供应商的 AI 经验水平。

机器学习算法应该对您的安全操作中心的各种专业人员（从经验丰富的分析师到新员工）有用，所以要确保您的供应商能够在每个级别上解决专业知识和知识缺口。

人工智能正在改变网络安全威胁的格局。保持消息畅通，并准备好使用它来防范其所构成的威胁。

# Text B

Bridging the divide between hype and reality when it comes to what artificial intelligence and machine learning can do to help protect a business.

While artificial intelligence and machine learning are far from new, many in security suddenly believe these technologies will transform their business and enable them to detect every cyber threat that comes their way. But instead, the hype may create more problems than it solves.

Recently, cybersecurity firm ESET surveyed 900 IT decision makers on their opinions of artificial intelligence and machine learning in cybersecurity practices.

According to the research, "the recent hype surrounding Artificial Intelligence (AI) and Machine Learning (ML) is deceiving three in four IT decision makers (75 percent) into believing the technologies are the 'silver bullet' to solving their cybersecurity challenges."

The hype, ESET says, causes *confusion* among IT teams and could put organizations at greater risk of falling *victim* to cybercrime. According to ESET's CTO, Juraj Malcho, "when it comes to AI and ML, the *terminology* used in some marketing materials can be misleading and IT decision makers across the world aren't sure what to believe."

Looking past the hype cycle, IT teams can achieve real value

| New Words and Expressions |
|---|
| **confusion**/kənˈfjuːʒ(ə)n/ n. |
| 混淆，混乱；困惑 |
| **victim**/ˈvɪktɪm/n. |
| 受害人；牺牲品；牺牲者 |
| **terminology**/ˌtɜːmɪˈnɒlədʒɪ/ n. |
| 术语，术语学；用辞 |

from machine learning and artificial intelligence *available* today.

## Types of Learning

Despite what marketing-speak says, there are different ways to implement machine learning–supervised or *unsupervised* learning.

In supervised learning, specific data is *collected* and defined output is used to create programs. This requires *actual* training of the system. In other words, a human must provide the expected output data to make the system useful. Most IT teams are *reluctant* to do this because it doesn't remove the human from the system.

Unsupervised learning is what the market is looking for, as it does remove the human. You don't need the output in this model. Instead, you feed data into the system and it looks for *patterns* from which to program *dynamically*.

## Ask the Right Questions

Most IT teams want to simply ask broad questions and get results to queries like, "find lateral movement[1]." Unfortunately, this is not possible today.

But you can use ML/AI to identify characteristics of lateral movement by asking questions like "Has this user logged in during this timeframe?" "Has the user ever connected to this server?" or "Does the user typically use this computer?" These types of questions are descriptive, not predictive. They infer answers by comparing new and historical data.

Analysts follow an attack down a logical path and ask questions at each step. Computers identify deviations from baselines and determine the risk level tracing the anomalies. This is the intersection where machines and humans come together for better results.

## What can be done today with ML/AI?

In reality, you must identify a strong baseline of the data structure to get value from ML/AI. Only then can you evaluate input data and make associations between the input data and the normal state of the network.

Here are threats that ML/AI can identify:

## DNS Data Exfiltration

While this is difficult to prevent, it is easily detected because the system can examine DNS traffic and know when DNS queries go to an authoritative server, but don't receive a *valid response*.

## *Credential* Misuse

According to Verizon's 2018 Data Breach *Investigations*

---

### New Words and Expressions

**available**/əˈveɪləb(ə)l/ adj.

可获得的；可购得的；可找到的；有空的

**unsupervised**/ˌʌnˈsjuːpəvaizd/ adj.

无人监督的；无人管理的

**collect**/kəˈlekt/ v.

收集；聚集；募捐

**actual**/ˈæktʃʊəl/ adj.

真实的，实际的；现行的，目前的

**reluctant**/rɪˈlʌkt(ə)nt/ adv.

不情愿的；勉强的；顽抗的

**pattern**/ˈpæt(ə)n/n.

模式；图案；样品

**dynamically**/daɪˈnæmɪkli/ adv.

动态地；充满活力地；不断变化地

**valid**/ˈvælɪd/ adj.

有效的；有根据的；合法的；正当的

**response**/rɪˈspɒns/ n.

响应；反应；回答

**credential**/krɪˈdenʃ(ə)l/ n.

证书；凭据；国书

**investigations**/ɪnˌvestɪˈɡeɪʃ(ə)n/ n.

调查；调查研究

report, humans are one of the biggest problems for organizations. Ninety-six percent of attacks come from email. On average only 4 percent of people fall for any given phishing attack, but a malicious actor only needs one victim to provide credentials.

Machine learning is useful here because the users have been baselined. Those users connect to and log in to a set number of devices each day. It's easy for a human to see when a credential is tried hundreds of times on a server, but it's hard to catch someone that tries to connect to 100 different machines on the network and only succeeds once.

While we are far from a type of artificial intelligence that can solve all cybersecurity problems, it is important to understand what's real and what's hype. As Malcho stated, "the reality of cybersecurity is that true AI does not yet exist. As the threat landscape becomes even more complex, we cannot afford to make things more confusing for businesses. There needs to be greater clarity as the hype is muddling the message for those making key decisions on how best to secure their company's networks and data."

Ultimately, the best solutions will be a combination of both supervised and unsupervised learning models: leveraging supervised learning to identify granular patterns of malicious behavior, while unsupervised algorithms develop a baseline for anomaly detection. Humans will not be eliminated from this equation any time soon.

## Terms

### Lateral Movement
攻击者进入目标网络后，下一步就是在内网中横向移动，然后再获取数据，所以攻击者需要一些立足点，因此横向移动会包含多种方式。

## 参考译文 B

当人们认识到人工智能和机器学习能够为企业提供安全保障时，就会弥补宣传概念与现实应用之间的鸿沟。虽然人工智能和机器学习不是什么新鲜事，但很多安全领域的人都相信这些技术将改变他们的业务，使他们能够检测到来自他们身边的网络威胁。相反，炒作可能带来的问题比它原本需要解决的问题还要多。

最近，网络安全公司 ESET 调查了 900 名 IT 决策者对网络安全实践中的人工智能和机器学习的看法。

关于人工智能（AI）和机器学习（ML）的炒作欺骗了四分之三（75%）的 IT 决策者，让他们相信这些技术是应对网络安全挑战的高招。

ESET 说，这种炒作会在 IT 团队中引起混乱，可能会使组织成为网络犯罪的受害者。ESET 的首席技术官 Juraj Malcho 说：“当涉及 AI 和 ML 时，一些营销材料中使用的术语可能会产生误导，全世界的 IT 决策者都不知道该相信什么。”

回顾炒作周期，IT 团队可以从机器学习和人工智能中获得真正的价值。

### 学习类型

不管市场营销怎么说，实现机器学习有不同的方法——监督学习和非监督学习。

在监督学习中，收集特定的数据用于创建程序，这需要对系统进行训练。换句话说，人类必须提供预期的输出数据。大多数 IT 团队不愿意这样做，因为它没有将人从系统中移除。

无监督学习正是市场所寻找的，因为不需要模型中的输出。

### 提出正确的问题

大多数 IT 团队只想简单地问一些宽泛的问题，然后得到诸如"查找横向移动"之类的查询结果。我们可以使用 ML/AI 来识别横向移动的特征，会有这样的问题："这个用户在这个时间段内登录了吗？""用户曾经连接到这个服务器吗？"或者"用户通常使用这台计算机吗？"这类问题是描述性的，而不是预测性的。他们通过比较新数据和历史数据来推断答案。

分析人员沿着一条合乎逻辑的路径跟踪攻击，并在每一步提出问题。计算机识别基线的偏差并确定跟踪异常的风险级别。这是机器和人类为了更好的结果走到一起的交叉点。

今天用 ML/AI 可以做什么？

实际上，您必须识别数据结构的一个强大基线，才能从 ML/AI 中获得值。只有这样才能对输入数据进行评估，并将输入数据与网络的正常状态进行关联。

以下是 ML/AI 可以识别的威胁：

### DNS 数据泄露

虽然这很难防止，但很容易检测到，因为系统可以检查 DNS 流量，并知道 DNS 查询何时进入服务器，但不会收到有效响应。

### 凭据滥用

根据 Verizon 2018 年的数据泄露调查报告，人类是最大的问题之一。96%的攻击来自电子邮件。平均只有 4%的人会被网络钓鱼攻击所欺骗。机器学习在这里很有用，因为用户已经被基线化了。这些用户每天连接并登录设备。当凭证在服务器上被尝试数百次时便很容易让人看到。

虽然我们远不是一种能够解决所有网络安全问题的人工智能，但了解什么是真实的，什么是炒作很重要。正如马尔乔所说，"网络安全的现实是，真正的人工智能还不存在。随着威胁形势变得更加复杂，我们不能让企业更困惑。对于那些在保护公司网络和数据安全方面做出关键决定的人来说，这种炒作混淆了信息。

最终，最好的解决方案是有监督和无监督学习模型相结合：利用有监督学习识别恶意行为的模式，而无监督算法是开发异常检测的基线，人类不会很快从这个系统模型中消失。

# Chapter *9*

# Artificial Intelligence in Finance

## Text A

The Financial *Stability* Board defines financial technology as "technologically enabled financial *innovations* that could result in new business models, *applications*, processes, or products with an associated material effect on financial markets and *institutions* and on the provision of financial services."

While innovation in finance is not a new concept, the focus on *technological* innovations and its pace have increased significantly. Fintech solutions that make use of big data *analytics*, artificial intelligence and *blockchain*[1] technologies are currently introduced at an unprecedented rate. These new technologies are changing the nature of the financial industry, creating many opportunities that offer a more *inclusive* access to financial services. The advantages notwithstanding, FinTech[2] solutions leave the door open to many risks, that may hamper consumer protection and financial stability. Relevant examples of such risks are *underestimation* of creditworthiness, market risk uncompliance, fraud *detection*, and cyber-attacks. Indeed fintech risk management represent a central point of interest for regulatory authorities, and require research and *development* of novel measurements.

Across the world, there is a strong need to improve the competitiveness of the fintech sector, introducing a risk management *framework* that can supervise fintech innovations without stifling their economic potential. A framework that can help both fintechs

---

**New Words and Expressions**

**stability**/stə'bɪlɪtɪ/ n.
　稳定性；坚定，恒心
**innovation**/ˌɪnəˈveɪʃn/ n.
　创新，革新；新方法
**application**/ˌæplɪˈkeɪʃ(ə)n/ n.
　应用；申请；应用程序；敷用
**institution**/ˌɪnstɪˈtjuːʃ(ə)n/ n.
　制度；建立；公共机构；习俗
**technological**/teknəˈlɒdʒɪk(ə)l/ adj.
　技术的；工艺的
**analytics**/ænəˈlɪtɪks/ n.
　分析学；解析学
**blockchain**/'blɒkˌtʃeɪn/ n.
　区块链
**inclusive**/ɪnˈkluːsɪv/ adj.
　包括的，包含的
**underestimation**
/ˈʌndəˌɛstɪˈmeɪʃən/ n.
　低估；过低的估价
**detection**/dɪˈtekʃ(ə)n/ n.
　侦查，探测；发觉，发现；察觉
**development**/dɪˈveləpm(ə)nt/ n.
　发展；开发；发育；住宅小区
**framework**/'freɪmwɜːk/ n.
　框架，骨架；结构，构架

and supervisors: On one hand, fintech firms need advice on how to identify opportunities for innovation procurements, for example in advanced regulatory technology (RegTech) solutions; On the other hand, the *supervisory* bodies' ability to monitor innovative financial products proposed by fintechs is limited, and advanced supervisory technology (SupTech) solutions are required. A crucial step in transforming compliance and supervision is to develop uniform and technology-driven risk management tools which could reduce the barriers between fintechs and supervisors.

We believe that a focused international research activity, coordinated at the level of a highly reputed open access scientific *journal* with multiple key *foci*, such as Frontiers in Artificial Intelligence, can help to close the gap between technical and *regulatory* expertise, in particular providing risk management procedures common to both sides. It could lead to the development of a regulatory framework that encourages innovations in big data analytics, artificial intelligence and blockchain technologies which, at the same time, satisfies supervisory concerns to apply *regulations* in an effective and efficient way and which protect consumers and investors.

Regulations and related supervisory requirements are placing great focus on risk management practices, which in turn drives the need for deep, transparent and auditable data analyses across organizations. Technologies such as big data analytics, artificial intelligence and blockchain ledgers may address risk management requirements and the associated costs more efficiently. In particular, these technologies can: (i) reduce credit scoring bias and improve fraud detection in peer-to-peer lending; (ii) measure and monitor *systemic* risk in peer-to-peer lending; (iii) measure and monitor market risk and volatility in financial markets (iv) enhance client risk profile matching in robo-advisory; (v) identify *illegal* activities in crypto markets, including fraudulent initial coin offerings and money laundering; (vi) identify and *prioritize* IT operational risks and cyber risks.

In line with these developments, the specialty section "Artificial Intelligence in Finance" of Frontiers in Artificial Intelligence aims to create an international research forum that provides and publishes key research on shared risk management solutions that *automatize compliance* of fintech companies (RegTech) and, at the same time,

---

### New Words and Expressions

**supervisory**/ˈsjuːpəˌvaɪzərɪ/ adj.
监督的

**journal**/ˈdʒɜːn(ə)l/ n.
日报，杂志；日记；分类账

**foci**/ˈfəʊkaɪ/ n.
焦点；焦距；聚焦

**regulatory**/ˈrɛɡ jʊˌleɪtəri/ adj.
管理的；控制的；调整的

**regulation**/regjʊˈleɪʃ(ə)n/ n.
管理；规则；校准

**systemic**/sɪˈstemɪk; / adj.
系统的；全身的；体系的

**illegal**/ɪˈliːg(ə)l/ adj.
非法的；违法的；违反规则的

**prioritize**/praɪˈɒrətaɪz / v.
给……排出优先级；优先处理；优先考虑

**automatize**/ɔːˈtɒmətaɪz/ v.
使自动化

increases the efficiency of supervisory activities (SupTech). The Artificial Intelligence in Finance section also builds synergies with the broader tech-focused specializations with its own journal and within Frontiers in Big Data and Frontiers in Blockchain.

Currently, *supervisors* and fintechs do not have a common framework to understand the opportunities/risks balance, leading to different perceptions. Artificial Intelligence in Finance aims to provide a research forum to discuss solutions that efficiently automatize both fintech compliance (RegTech) and supervisory monitoring (SupTech).

The vision of Artificial Intelligence in Finance is to build a *collaborative* innovative environment from which both supervisory bodies and regulated institutions can benefit. *Specifically*, we aim at connecting the two sides of the coin by organizing a forum for research discussion which will have the purpose of sharing risk *measurement* solutions that fit the needs of both regulated institutions and regulators. The discussion will draw on the *contribution* from three types of project participants:

- Fintech and financial companies, who have a detailed understanding of business models based on *financial* technologies;
- Regulators and supervisors, who have a detailed understanding of the regulations and risks that concern *financial* technologies;
- Universities and research centers, which have a detailed understanding of the risk *management* models that can be applied to financial technologies.

Conceptually, the research content of the journal will be classified around three types of FinTech risk management models, which will constitute the conceptual map of the journal The classification is based on the three main technologies that drive FinTech innovations:

- Big data analytics, with its application to peer-to-peer lending, with main risks arising from credit risk, and systemic risk;
- Artificial intelligence, with its application to financial robo-advice, with main risks arising from market risk and compliance risk;
- Blockchain, with main application to crypto-assets, with main risks arising from fraud detection, money laundering risk, IT operational risk and cyber risks.

**New Words and Expressions**

**compliance**/kəmˈplaɪəns / n.
顺从，服从；承诺
**supervisor**/ˈsuːpəvaɪzə/ n.
监督人，指导者；管理人；检查员
**collaborative**/kəˈlæbəretɪv/ adj.
合作的，协作的
**specifically**/spɪˈsɪfɪkəlɪ / adv.
特别地；明确地
**measurement**/ˈmeʒəm(ə)nt/ n.
测量；度量；尺寸；量度制
**contribution**/kɒntrɪˈbjuːʃ(ə)n/ n.
贡献；捐献；投稿
**financial**/faɪˈnænʃ(ə)l/ adj.
金融的；财政的，财务的
**management**/ˈmænɪdʒm(ə)nt/ n.
管理；管理人员；管理部门；操纵

Artificial Intelligence in Finance will consider research from all the above three areas. Research in Big data and Blockchain, but also in AI more generally, neatly connects to other Frontiers' journals, such as Frontiers in Big Data and Frontiers in Blockchain. This *interdisciplinary* infrastructure aims to leverage collaborations and the expertise of diverse research communities—something that is at the center of FinTech innovation.

### Innovative Technologies

The European Commission argues that the term big data refers to "large amounts of different types of data produced with high velocity from a high number of various types of sources." Big data analytics refers to the variety of technologies, models and procedures that involve the analysis of big data aimed revealing *insights*, patterns of *causality* and of correlation, and to predict future events.

Over the years, academics and experts in computer science and statistics have developed advanced techniques to obtain insights from large datasets combining a variety of data types obtained from a variety of sources. These models are able to utilize the ability of computers to perform complicated tasks by learning from experience. Following a definition offered by the Financial Stability Board artificial intelligence is a broad term capturing "the application of *computational* tools to address tasks traditionally requiring human sophistication." It is important to mention that often the terms AI and machine learning are used *interchangeably*. However, Artificial Intelligence is a broader term, of which machine learning represents a subcategory: the difference being that machine learning is a data-driven way to achieve AI, but not the only one; similarly, big data analytics is broader then machine learning, as it includes also statistical learning. For a further discussion on the difference between AI and Machine Learning, see also Kersting.

Among the emerging technologies with *significant* potential to change the financial systems and industry from its core, the blockchain has received a significant amount of attention over the last few years. A blockchain is a distributed database of records of all *transactions* or digital events that have been executed and shared among participating parties. Each transaction in the distributed database of records is verified by the participants through a majority

**New Words and Expressions**

**interdisciplinary**
/ˌɪntəˈdɪsɪplɪn(ə)rɪ/ adj.
各学科间的；跨学科的

**insight**/ˈɪnsaɪt/ n.
洞察力；洞悉

**causality**/kɔːˈzælɪtɪ/ n.
因果关系

**computational**
/ˌkɒmpjʊˈteɪʃənl/ adj.
计算的

**interchangeably**
/ˌɪntəˈtʃendʒəbli / adv.
可交换地

**significant**/sɪgˈnɪfɪk(ə)nt/ n.
象征；有意义的事物

**transaction**/trænˈzækʃ(ə)n / n.
交易；事务；办理；会报，学报

consensus and, once confirmed, the transaction can never be altered or deleted. Hence, the blockchain contains a certain and verifiable record of every single transaction ever made between the participants in a network.

## Financial Applications

Many fintech applications rely on big data analytics and, in particular, those based on peer-to-peer (P2P) financial transactions, such as peer to peer lending, *crowdfunding*, and invoice trading. The concept peer-to-peer captures the interaction between units, which eliminates the need for a central intermediary. In particular, peer-to-peer lending enacts disintermediation by allowing borrowers and lenders to communicate directly, using the platform as an information provider which, among other things, assesses the credit risk of borrowers. From a regulatory perspective, a key point of interest is whether such credit risk measurements reflect the actual capacity of borrowers to repay their debt. Regulation must be technologically neutral and, therefore, credit risk compliance should be imposed on fintechs as they are for banks. At the same time, it cannot be so burdensome to disincentives the growth of alternative financial service provides.

Automated consultants, known as robot advisors, are considered the main application of AI in financial services. The European Supervisory Authorities joint report defines the phenomenon of automation in financial advice as "a procedure in which advice is provided to consumers without, or with very little human intervention and with providers relying on computer-based algorithm and/or decision trees." In practice, robot advisors build personalized portfolios for investors, on the basis of algorithms that take into account investors' information such as age, risk tolerance and aversion, net income, family status. Obtaining this information is a legal requirement and robot advisors employ online questionnaires to obtain it.

Crypto assets are the main application of blockchain technology and are considered one of the largest markets in the world which remain unregulated. Within the last decade, digital currencies, operating independently of central banks have massively grown in popularity, price, and volatility. The Bitcoin is the oldest, most popular and widely used digital currency, and it offers low-cost, decentralized transfer of value anywhere in the world with the only *constraint*

*New Words and Expressions*

**crowdfunding**/kraʊdˈfʌndɪŋ/ n.
众筹

**constraint**/kənˈstreɪnt/ n.
约束；局促，态度不自然；强制

representing the availability of an internet connection. However, many other crypto assets are available, and new ones are continuously emerging through Initial Coin Offerings, in which a company sells digital tokens that *eventually* can be exchanged for goods, services or other currencies. It is a new fundraising *method*, which combines elements of both crowdfunding and traditional initial public offerings.

### Risk Concerns and Management

Although there are many existing legislations that are intended to serve in the interest of consumer and investor protection, lending fintechs give rise to "disintermediation," which requires the need for further protection of consumers and investors. In the case of peer to peer lending, there are two main causes of concern. First, P2P *platforms* have less information on their borrowers, compared to classical banks, and are less able to deal with asymmetric information. Second, in most P2P lending platforms the credit risk is not held by the platform but, rather, by the investors. Both causes lead to a high likelihood that the scoring system of P2P lenders may not adequately reflect the "correct" probability of default of a loan. A further issue associated with the nature of P2P platforms is that they give rise by construction to globally interconnected networks of transactions. This suggests that they cannot avoid the measurement of systemic risks arising from contagion mechanisms between borrowers.

In the context of P2P lending, a key risk to measure is the risk associated with the *default* of borrowers: credit risk. Statistical theory offers a great variety of supervised models for credit scoring and credit risk management and, in particular, logistic regression, and generalized linear models. The same models can be applied to similar classification problems in peer-to-peer lending, such as consumer's fraud and money laundering detection.

A key issue that arises in employing generalized linear models for P2P classification problems is that the event to be predicted is multivariate. To solve this issue, Lauritzen introduced graphical models to model dependencies between random variables, by means of a unifying and powerful concept of a mapping between probabilistic conditional independences, missing edges in a graphical representation, and suitable statistical model parameterization. In

***New Words and Expressions***

**eventually**/ɪˈventʃʊəlɪ / adv.

最后，终于

**method**/ˈmeθəd/ n.

方法；条理；类函数

**platform**/ˈplætfɔːm/ n.

平台；月台，站台；坛；讲台

**default**/dɪˈfɔːlt/ n.

违约；缺席；缺乏；系统默认值

parallel, Mantegna introduced hierarchical structures in financial markets, based on correlation matrices, developing a powerful distance-based statistical model able to uncover similarity relationships among financial assets. These models have been applied in a variety of financial contexts, including credit scoring, churn modeling, and fraud detection.

In line with these developments, Giudici and Hadji-Misheva suggest to model credit risk of peer to peer lending taking advantage of their natural interconnectedness, by means of *correlation* network models, a subset of *graphical* models that has been introduced in finance to measure systemic risks risk. This allows to improve the accuracy of credit risk models and, furthermore, to measure a risk type that is particularly evident in P2P lending: systemic risk, recently applied to bank, and sovereign default. Giudici and Hadji-Misheva show how to build a correlation network for P2P lending: associating each borrower with a statistical unit, at each time point many variables can be observed for that unit; in the case of SME lending, balance sheet variables; in the case of consumer credit, transaction account variables. A correlation network between borrowers can then be built on the basis of the observed values of one variable over time. Associating each borrower with a node in the network, each pair of nodes can be thought to be connected by an edge, whose weight is equal to the correlation coefficient between the two-time series of the chosen variable, each corresponding to a specific borrower. If we consider all pairs of borrowers, we will get a matrix of correlation weights, also known as "adjacency matrix." Once the adjacency matrix is derived, summary network centrality measures suggest which are the most important units in the network or, in financial risk terms, which are the most contagious borrowers show, in real P2P lending data analysis that, when network centrality measures are embedded in a generalized linear model specification, they can improve the predictive accuracy of credit scoring algorithms.

Moving to asset management fintechs, note that the advantages associated with automatized advice may be offset by the greater risks that are brought on board, among which the risks of making unsuitable decision and risks of errors and functional limitations of the tool. As is the case with big data analytics, there are several regulatory requirements that already exist and apply to automated

**New Words and Expressions**

**correlation**/ˌkɒrəˈleɪʃ(ə)n/ n.
相关，关联；相互关系
**graphical**/ˈɡræfɪk(ə)l/ adj.
图解的；绘画的；生动的

advice. However, some risks are yet to be fully considered and measured. Among them, we believe the following are the most relevant: (i) compliance risk—mismatch between expected and actual investment risk class; (ii) market risk—the likelihood that adverse movements and volatility in financial markets, either traditional or new cause unexpected losses in investors' portfolios.

As for peer to peer lending, the increased risks connected with the use of robot advisory platforms can be mitigated by an appropriate analysis of the data they generate. In this respect, robot advisors generate, in an automated way, a large amount of data, which can be leveraged not only to improve the service, making it more personalized, but also to reduce compliance risk and, in particular, the risk of an incorrect profile matching between "expected" and "actual" risk classes.

Recent studies have shown that an *accurate* analysis of risk propensity *questionnaires* can allow robo-advisors to estimate the "expected" risk class of each investor. Data analysis algorithms can be implemented also on the supply side, considering the returns of the available financial products to classify them into homogeneous "actual" risk classes. Linking together the "expected" risk classification of an investor with its "actual" classification allows to evaluate whether a robot advisor respects its risk profile, one of the most important requirement of the MIFID regulation, which thus becomes, in the context of robot advisory, a verifiable requirement, not only from a formal viewpoint, but also from an operational one.

The literature on the measurement of expected risks in robot advisory is very limited. Scherer investigates, within a machine learning approach based on tree models, the key investor characteristics that can predict financial market participation; Alexy et al. is a related work. Similarly, the literature on the measurement of the actual risk of a given set of financial products is also very limited. Tumminello et al. and Tola et al., who employ clustering models to construct homogeneous asset classes, and who considered interconnectedness risk, are noticeable exceptions.

Giudici and Polinesi extend Scherer's approach deriving expected risk classes from the responses to the MIFID questionnaire, building correspondence analysis models on the observed contingency table, that results from the cross-classification of the responses to the

**New Words and Expressions**

**accurate**/ˈækjərət/ adj.
精确的
**questionnaire**/ˌkwestʃəˈneə(r)/ n.
问卷；调查表

questionnaire. They also show how to employ feed forward neural network models to estimate the risk class of a given investor's portfolio, on the basis of the observed returns. By comparing the expected with the actual risk class, for a sample of investors, it is thus possible to evaluate, in an automated way, whether the robot advisor is compliant with the risk profile of the investor.

We remark that specific concerns arise, from a market risk viewpoint, when crypto assets are combined with classical ones in investment activities. In particular, bitcoins, and crytpoassets have been associated with exceptionally high *volatility* and greatly sensitive prices. Indeed, fluctuations are very common throughout the existence of the crypto assets, which in turn raises the question whether this *behavior* is attributed to general market conditions or to idiosyncratic factors. To address these concerns, network models take the central stage, as could be expected. Nakamoto described the bitcoin as a purely peer-to-peer version of electronic cash that allows online payments to be send directly from one party to another, without going through a financial institution. Hence, in its essence, the bitcoin represents a solution to the double-spending problem using a peer-to-peer network. This suggests that a correct measure of the risks associated with this technology must take into account the interconnections generated by network transactions.

In this context, correlation network models can be employed to detect the main determinants of volatility. More recently, have applied correlation VAR models to check whether price *contagion* between different bitcoin exchange markets exist, and found that this is the case, especially for smaller exchanges.

Many innovative fintechs have payment deals with the application of blockchain technology. The main risk concerns about blockchain applications in finance relate to operational risks. Many international regulatory authorities have raised significant concerns suggesting that, in most cases, small investors do not adequately understand the risk involved with Initial Coin Offerings. Although many legitimate start-ups use ICOs for the purpose of raising money, many projects also exist which do not intend to deliver any value to the investors. The market has seen many such cases of fraudulent ICOs which raises deep concerns for investor protection and overall financial stability. To identify the main determinants of

**New Words and Expressions**

**volatility**/ˌvɒləˈtɪlətɪ / n.

挥发性；易变；活泼

**behavior**/bɪˈheɪvjə/ n.

行为，举止；态度；反应

**contagion**/kənˈteɪdʒ(ə)n/ n.

传染病；蔓延；触染

fraudulent ICOs, text mining analytics methods, that use network models to reduce their curse of dimensionality, can be applied. Following most recent statistics, 99% of all ICOs use Telegram as a channel for interacting with their communities. Typically, the Telegram groups are characterized by many members and detailed discussions about the value of the individual projects, as well as, by the expectations of the communities concerning the success of the ICO and the company. By collecting data from the Telegram ICOs and discussions on Telegram chats relating the value and prospects of the projects in question, we can build, train, and test supervised models to discriminate and classify ICOs by their probability of fraud, using for example the methods shown in Hochreiter.

　　Another cause of concern is that crypto assets allow for a multi-billion dollar global market of anonymous transactions, which does not undergo any control. Hence, its emergence and growth can create considerable challenges for market *integrity*, particularly from money laundering activities. Money laundering embraces all those operations to disguise the illicit origin of capital, to give it a semblance of legitimacy, and facilitate the subsequent reinvestment in the lawful economy. A recent study conducted by Foley et al. aims at quantifying and characterizing the illegal trade facilitated by the Bitcoin, to provide a better understanding of the nature and scale of the problem facing this technology. The results from the study suggest that *approximately* one-quarter of Bitcoin users and one-half of Bitcoin transactions are associated with illegal activity. The authors found that around $72 billion of illegal activity per year involves Bitcoin, which is close to the scale of the US and European markets for illegal drugs. In the context of money laundering detection, network-based community detection models can be employed. They exploit the transactional network topology for the purpose of identifying communities of users and, in particular, to identify communities of money launderers, using the transactions between them. More formally, the method that can be applied is a network cluster analysis *algorithm* that takes as inputs the set of users and the trades between users. The output of the algorithm is an assign-ment of users to communities such that the "modularity" of the communities is maximized.

　　An additional cause of concern is that cryptoassets are fully

---

**New Words and Expressions**

**integrity**/ɪnˈtegrɪtɪ/ n.

完整；正直；诚实；廉正

**approximately**/əˈprɒksɪmətlɪ/ adv.

大约，近似地；近于

**algorithm**/ˈælgərɪð(ə)m/ n.

算法，运算法则

digital and, therefore, may lead to higher IT operational risks, such as errors in the functioning of the algorithms, and hacking and manipulation of the algorithms, to name only a few. While the literature on the quantitative measurement of operational risk constitute a reasonably large body, that on cyber risk measurement is very limited. As cyber risks are very different in nature, rare, and typically not repeatable, a useful approach to measure them is to consider an ordinal-based, scorecard approach, similar to that done in self- assessment-based operational risk management, in reputation measurement or in portfolio analysis using stochastic dominance.

In this way a cyber risk measure can be used to rank cyber risks and prioritize interventions, preventing failures and reducing ex-ante the impact of risks. This on the basis of ordinal random variables, that represent the levels of frequency and severity for different cyber risk events, in different business areas. A similar approach can be consistently undertaken to measure operational risks deriving from the use of robo-advisors, caused by their malfunctioning, rather than by cyber-attacks. Note also that an ordinal-based measurement of operational risks and cyber risks can be easily adapted to scenario testing, which is one of the best ways for the financial industry to protect from them, specifically when they are conducted across the industry.

## Terms

### 1. Blockchain

一般认为，区块链是一种按照时间顺序将数据区块以链条的方式组合形成的特定数据结构，并以密码学方式保证其不可篡改和不可伪造的去中心化、去信任的分布式共享总账系统。

### 2. FinTech

FinTech 是 Financial Technology 的缩写，可以简单理解为 Finance（金融）+Technology（科技），指通过利用各类科技手段创新传统金融行业所提供的产品和服务，提升效率并有效降低运营成本。

## Comprehension

### Blank Filling

1. Fintech solutions that make use of _____ , _____ and _____ technologies are currently introduced at an unprecedented rate.

2. Technologies such as _____, _____ and _____ ledgers may address risk management requirements and the associated costs more efficiently.

### Content Questions

1. What is financial technology?

2. What is the vision of Artificial Intelligence in Finance?

3. What are fintech applications?

## Answers

**Blank Filling**

1. big data analytics; artificial intelligence; blockchain
2. big data analytics; artificial intelligence; blockchain

**Content Questions**

1. The Financial Stability Board defines financial technology as "technologically enabled financial innovations that could result in new business models, applications, processes, or products with an associated material effect on financial markets and institutions and on the provision of financial services."

2. The vision of Artificial Intelligence in Finance is to build a collaborative innovative environment from which both supervisory bodies and regulated institutions can benefit. Specifically, we aim at connecting the two sides of the coin by organizing a forum for research discussion which will have the purpose of sharing risk measurement solutions that fit the needs of both regulated institutions and regulators.

3. Many fintech applications rely on big data analytics and, in particular, those based on peer-to-peer (P2P) financial transactions, such as peer to peer lending, crowdfunding, and invoice trading.

## 参考译文 A

金融稳定委员会将金融技术定义为技术支持的金融创新，可以产生新的商业模式、应用程序、流程或产品，并对金融市场和机构以及金融服务的提供产生相关的实质性影响。

虽然金融创新并不是一个新概念，但对技术创新的关注和创新步伐已经显著提高。利用大数据分析、人工智能和区块链技术的金融科技解决方案目前正以前所未有的速度推进。这些新技术正在改变金融行业的性质，创造了许多机会，使金融服务的获取更具包容性。尽管有这些优势，但金融科技也为许多风险敞开了大门，这些风险可能会妨碍消费者的金融稳定。这类风险包括对信誉的低估、欺诈检测和网络攻击。实际上，金融科技风险管理是监管机构关注的中心问题，需要研发新的衡量标准。

世界各地都迫切需要提高金融技术部门的竞争力，引入一个既能监督金融技术创新，又不会抑制其经济潜力的风险管理框架，一方面，金融技术公司需要关于如何识别创新采购机会的建议，例如在先进监管技术（RegTech）解决方案中；另一方面，监管机构对金融技术创新产品的监管能力有限，需要先进的监督技术（SupTech）解决方案。转变合规和监管的一个关键步骤，是开发统一的、技术驱动的风险管理工具，以减少金融技术人员和监管人员之间的障碍。

我们相信国际研究活动主要协调获取科学期刊关键焦点，如人工智能前沿，可以帮助缩小技术和管理知识之间的差距，特别是双方共同提供风险管理过程。它可能会催生一个监管框架，鼓励在大数据分析、人工智能和区块链技术方面的创新，同时满足监管要求，以有效和高效的方式实施监管，保护消费者和投资者。

规章制度和相关的监督要求非常重视风险管理实践，而风险管理实践反过来又推动了对跨组织的深入、透明和可审计数据分析的需求。大数据分析、人工智能、区块链账簿等技术可以更有效地满足风险管理需求和相关成本。具体来说，这些技术可以：①减少信用评分偏差，提高 P2P 借贷的欺诈检测能力；②衡量和监测个人对个人贷款的系统风险；③衡量和监测金融市场的市场风险和波动情况；④加强智能咨询客户风险配置匹配；⑤查明加密市场的非法活动，包括伪造首次硬币和洗钱；⑥识别和区分 IT 运营风险和网络风险。

人工智能的目标是创造一个国际研究论坛，提供共享和发布关键研究风险管理解决方案，使金融科技（Fintech）公司提高管理（RegupTech）活动的效率，同时提高监督（SupTech）活动的效率。金融领域的人工智能部门还与更广泛的以技术为重点的专业领域、专属期刊以及大数据领域和区块链领域的前沿领域建立了协同效应。

目前，金融科技没有框架来理解机会/风险的平衡，导致不同的看法。金融领域的人工智能旨在提供一个研究论坛，讨论金融技术合规（RegTech）和监督监控（SupTech）高效自动化的解决方案。

人工智能在金融领域的愿景是建立一个协作创新的环境，监管机构和受监管机构都能从中受益。具体来说，我们的目标是通过组织一个论坛来进行研究讨论。这个论坛的目的是分享符合监管机构需求的风险度量解决方案。讨论将利用三种参与者的贡献：

- 金融科技和金融公司，对基于金融技术的商业模式有详细的了解；
- 对涉及金融技术的规章制度和风险有详细了解的监管人员；
- 大学和研究中心，他们对可以应用于金融技术的风险管理模型有详细的了解。

从概念上讲，期刊的研究内容将围绕三种金融技术风险管理模型进行分类，这三种模型构成了《华尔街日报》的概念图，基于推动金融技术创新的三种主要技术分类：

- 大数据分析应用于 P2P 借贷，主要风险来自信用风险和系统风险；
- 人工智能应用于金融机器人咨询，主要风险来源于市场风险和合规风险；
- 区块链主要应用于加密资产，主要风险来自欺诈检测、洗钱风险、IT 运营风险和网络风险。

金融领域的人工智能将考虑以上三个领域的研究。大数据和区块链的研究，以及更一般的人工智能的研究，连接到其他前沿期刊上，例如大数据前沿和区块链前沿。这种跨学科的基础设施旨在利用合作和不同研究团体的专业知识——这是金融技术创新的核心。

**创新技术**

欧盟委员会认为"大数据"一词指的是"大量不同类型的数据以高速度从大量不同类型的来源产生"。大数据分析是指对大数据进行分析的各种技术、模型和程序，其目的是揭示洞察、因果关系和相关性的模式，并预测未来的事件。

多年来，计算机科学和统计学方面的学者和专家开发了先进的技术，从大型数据集中获取见解，这些数据集结合了从各种来源获得的各种数据类型。这些模型能够从经验中学习利用计算机执行复杂任务的能力。根据金融稳定委员会给出的定义，人工智能是一个宽泛的术语，它捕捉"计算工具的应用，以解决传统上需要人类复杂性的任务"。很重要的一点是，人工智能和机器学习这两个术语经常可以互换使用。然而，人工智能是一个更广泛的术语，其中机器学习代表了一个子类：机器学习是实现人工智能的一种数据驱动方式，但不是唯一的一种；同样，大数据分析比机器学习更广泛，因为它还包括统计学习。关于人工智能和机器学习区别的进一步讨论。

在新兴技术中，区块链具有从核心向金融系统和行业转变的巨大潜力，在过去几年中，它受到了大量关注。区块链是一个分布式数据库，记录所有交易或数字事件，已执行和共享的参与各方。分布式记录数据库中的每一笔交易都由参与者通过多数人的一致意见进行验证，一旦被确认，该交易就不会被更改或删除。因此，区块链包含了网络参与者之间每一笔交易可验证的记录。

### 金融应用程序

许多金融技术应用程序依赖于大数据分析，尤其是基于 P2P 金融交易的应用程序，如 P2P 借贷、众筹和发票交易。尤其值得一提的是，点对点贷款通过允许借款人和贷款人直接沟通来实现非中介化，使用平台作为信息提供者，评估借款人的信用风险。从监管角度看，一个关键的利益点是，这种信用风险衡量是否反映了借款人偿还债务的实际能力。监管必须在技术上是中立的，因此，应该像对银行一样，对金融技术企业实行信贷风险合规监管。

被称为机器人顾问的自动化顾问被认为是人工智能在金融服务中的主要应用。欧洲监管当局联合报告将金融咨询中的自动化现象定义为"一种向消费者提供咨询的程序，在这种程序中，不需要或几乎不需要人工干预，而是由提供者依赖基于计算机的算法和/或决策树"。在实践中，机器人顾问基于考虑投资者年龄、风险承受力、净收益、家庭状况等信息的算法，为投资者构建个性化的投资组合。获取这些信息是一项法律要求，机器人顾问使用在线问卷来获取这些信息。

加密货币是区块链技术的主要应用，被认为是世界上最大的市场之一，目前仍不受监管。在过去的十年里，独立于央行运作的数字货币在受欢迎程度、价格和波动性方面都有了大幅度的增长。比特币是最古老、最受欢迎、使用最广泛的数字货币，它可在世界任何地方提供低成本、分散的价值转移，唯一的限制是互联网连接的可用性。然而，还有许多其他加密资产可供使用，而且通过首次发行硬币，新的加密货币不断涌现。在首次发行硬币的过程中，一家公司出售数字令牌，这些令牌最终可以用来交换商品、服务或其他货币。这是一种传统首次公开募股为一体的新型融资方式。

### 风险关注及管理

虽然现有的许多立法都是为了保护消费者和投资者的利益，但贷款金融技术会导致"脱媒"，这就需要进一步保护消费者和投资者。在对等借贷的情况下，有两个主要原因引起了关注。首先，与传统银行相比，P2P 平台的借款人信息更少，处理信息不对称的能力也更弱。其次，在大多数 P2P 借贷平台中，信用风险并不由平台承担，而是由投资者承担。与 P2P 平台性质相关的另一个问题是，它们通过构建全球互联的交易网络而产生。这表明，它们无法避免衡量借款人之间传染机制所引发的系统性风险。

在 P2P 借贷的背景下，衡量与借款人违约相关的风险：信用风险。统计理论为信用评分和信用风险管理提供了多种监督模型，特别是逻辑回归模型和广义线性模型。同样的模型也适用于 P2P 借贷中类似的分类问题，如消费者欺诈、洗钱检测等。

P2P 分类问题的广义线性模型中出现的一个关键问题是，要预测的事件是多元的。为了解决这个问题，Lauritzen 将图形模型引入随机变量之间的依赖关系模型中，通过联结函数建立响应变量的数学期望值与线性组合的预测变量之间的关系。同时，Mantegna 在金融市场中引入了基于相关矩阵的层次结构，建立了一个强大的基于距离的统计模型，能够发现金融资产之间的相似性关系。这些模型已被应用于各种金融环境，包括信用评分、客户流失建模和欺诈检测。

随着这些技术的发展，Giudici Hadji-Misheva 建议信用风险模型的点对点借贷需要利用其自身互联的属性，通过相关网络模型和一个子集的图形模型，来评估财务风险衡量系统性风险。这有助于提高信用风险模型的准确性，此外，还可以衡量 P2P 贷款中特别明显的一种风险类型：最近应用于银行的系统性风险和主权违约。Giudici 和 Hadji-Misheva 展示了如何构建 P2P 借贷的关联网络：将每个借款人与一个统计单元相关联，在每个时间点可以观察到该单元的多个变量；就中小企业贷款而言，观察资产负债表变量；在消费信贷方面，观察交易账户表变量。借款人之间的相关网络可以建立在一个变量随时间变化的观测值的基础上。将每个借方与网络中的一个节点关联起来，每对节点都可以认为是由一条边连接起来的，其权值等于所选变量的两时间序列之间的相关系数，每个对应于一个特定的借方。如果我们考虑所有借款人，我们将得到一个相关权矩阵，也称为"邻接矩阵"。一旦得到邻接矩阵，总结哪些是网络中最重要的单元，或者在金融风险方面，哪些是最具传染性的借款人，在真实 P2P 借贷数据分析中，将网络中心度量嵌入广义线性模型规范中时，可以提高信用评分算法的预测精度。

资产管理与自动化的建议可能会抵消相关的优势带来的更大的风险。正如大数据分析的情况一样，已经存在一些适用于自动建议的监管要求。然而，一些风险仍有待充分考虑和衡量。其中，我们认为最相关的是：①预期与实际投资风险等级的合规风险错配；②市场风险——金融市场的不利变动和波动，无论是传统的还是新的，都可能导致投资者投资组合的意外损失。

对于 P2P 借贷，通过对机器人咨询平台产生的数据进行适当的分析，可以降低与使用机器人咨询平台相关的风险。在这方面,机器人通过一个自动化的方式，这不仅可以利用改善服务，使它更个性化，而且减少合规风险。

关于机器人咨询中预期风险度量的文献非常有限。在基于树形模型的机器学习方法中，研究了能够预测金融市场参与的关键投资者特征是一项相关工作。同样，关于衡量特定金融产品的实际风险的文献也非常有限。

Giudici 和 Polinesi 扩展了 Scherer 的方法，从对 MIFID 问卷的响应中推导出预期风险等级，在观察到的列联表上建立对应分析模型，这是对问卷响应进行交叉分类的结果。它们还展示了如何使用前馈神经网络模型，以观察到的回报为基础，给投资者评估出风险等级。通过将预期风险与实际风险等级进行比较，对于一个投资者，就有可能以一种自动化的方式评估机器人顾问是否符合投资者。

我们注意到，从市场风险的观点来看，当加密资产与投资活动中的经典资产相结合时，会产生一些具体的担忧。在加密资产存在的整个过程中，波动非常普遍，这反过来又提出了一个问题，即这种行为是归因于一般市场条件还是特殊因素。为了解决这些问题，正如预期的那样，网络模型占据了中心舞台。Nakamoto 将比特币描述为一种纯粹的点对点电子货币，允许在线支付直接从一方发送到另一方，而无须通过金融机构。因此，从本质上说，比特币代表了一个使用点对点网络的双消费问题的解决方案。这表明，要正确衡量与这项技术相关的风险，必须考虑网络事务所产生的互连。

在这种背景下，相关网络模型可以用来检测波动率是主要决定因素。最近，应用相关VAR 模型检验不同比特币交易市场之间是否存在价格影响，发现情况确实如此，尤其是对于较小的交易所。

许多创新的金融技术都与区块链技术的应用有支付协议。区块链在金融领域应用的主要风险与操作风险有关。许多国际监管机构提出了重大担忧，表明在多数情况下，小投资者没有充分了解首次发行硬币的风险。虽然许多合法的初创公司使用 ICO 来筹集资金，但也有许多项目并不打算为投资者带来任何价值。市场上已经出现了许多这样的欺诈 ICO 案例，这引起了对投资者保护和整体金融稳定的深切关注。为了识别欺诈 ICO 的主要决定因素，可以使用文本挖掘分析方法。根据最近的统计数据，99%的 ICO 使用 Telegram 作为与社区互动的渠道。典型地，Telegram 小组的特点是有很多成员，详细讨论单个项目的价值，以及社区对 ICO 和公司成功的期望。通过收集的数据和讨论有关前景的项目，我们可以构建和测试监督模型。

一个令人担忧的原因是，加密资产允许数十亿美元的匿名交易。因此，它的出现和增长可能对市场的完整性造成相当大的挑战，特别是洗钱活动，包括所有那些掩盖资本非法来源的活动，使其具有表面上的合法性，并促进随后对合法经济的再投资。Foley 等人最近进行的一项研究旨在对比特币促成的非法交易进行量化和定性，以便更好地理解这种技术面临的问题的性质和规模。研究结果显示，大约四分之一的比特币用户和一半的比特币交易与非法活动有关。两位作者发现，每年大约有 720 亿美元的非法活动涉及比特币，接近美国和欧洲非法毒品市场的规模。在洗钱检测的背景下，可以采用基于网络的社区检测模型。它们利用事务性网络拓扑来标识用户的社区，特别是使用用户之间的事务标识洗钱者的社区。更正式地说，可以应用的方法是一种网络聚类分析算法，它将用户集和用户之间的交易作为输入。算法的输出是将用户分配给社区，使社区的"模块化"最大化。

另一个令人担忧的原因是，加密资产是完全数字化的，因此可能会导致更高的 IT 操作风险，例如算法运行中的错误，以及对算法的黑客和操纵。虽然关于操作风险定量度量的文献构成了一个相当大的主体，但是关于网络风险度量的文献非常有限。网络风险在本质上是截然不同的，而且通常是不可重复的。因此，衡量这些风险的一种有效方法是用一种基于顺序的记分卡方法，类似于基于自我评价的操作风险管理、声誉测量或使用随机优势的投资组合分析中的方法。

通过这种方式，可以利用网络风险测度对网络风险进行排序，对干预措施进行优先排序，防止失败，降低风险的影响。这是基于序号随机变量得出的，序号随机变量表示不同业务领域不同网络风险事件发生的频率和严重程度。我们可以采用类似的方法来衡量由于使用机器人顾问而产生的操作风险，这些顾问是由其故障而非网络攻击造成的。还要注意，基于顺序的操作风险和网络风险度量可以很容易地适用于场景测试，这是金融业在进行这些风险测试时保护自己不受影响的最佳方法之一。

# Text B

Figure 9-1

**New Words and Expressions**

**technology**/tek'nɒlədʒɪ/ n.
技术；工艺；术语

**function**/'fʌŋ(k)ʃ(ə)n/ n.
平功能；函数；职责

**neural**/'njʊər(ə)l / adj.
神经的；神经系统的；背的；神
经中枢的

**solution**/sə'luːʃ(ə)n/ n.
解决方案；溶液；溶解；解答

**computational**
/kɒmpjʊ'teɪʃənl / adj.
计算的

**organization**/ˌɔːgənaɪ'zeɪʃn/ n.
组织；机构；体制；团体

Artificial Intelligence (AI) is the emerging *technology* in the finance *function*. What are the opportunities?

## AI in Finance

AI itself is an encompassing term that embraces a number of technological advances including:

Machine learning — using *neural* networks, statistics and operational research to identify insights in data without being programmed what to conclude

Deep learning — using many layers of computing power and improved training techniques to identify patterns in data.

In delivering an AI *solution* you may be combining a number of technologies. For example, enhancing analytics and forecasts by using structured and unstructured data to deliver forward thinking insights. Cloud based storage may be an asset for the data volumes involved and the *computational* power needed.

*Organizations* have examples of using forms of AI to address some of the data validation errors encountered in RPA processes by applying machine learning to the errors. Some organizations are creating process chains where data is captured through chatbots, entered using RPA tools and errors resolved using machine learning.

## Implementing AI in finance

There are opportunities to apply AI in finance. For many organizations this is the level of opportunity. One which may yet only be emerging as a proof of concept rather than as a robust solution.

Perhaps the most significant issue for finance is the skills needed to support this next technological wave. AI needs individuals who understand the data and the processing capability and how to frame the problem. This takes the skills needed in finance to another level. Bias is an issue which these individuals need to be able to *recognise* and address. Bias in output occurs when the input data set is not representative of the outcome and the machine "learns" in the wrong way. This needs to be identified and corrected before reliance is placed upon the solution.

> **New Words and Expressions**
>
> **recognize**/ˈrekəgnaɪz/ v.
>
> 认出；承认，认可；识别

## 参考译文 B

人工智能（AI）是金融功能中的新兴技术。机会是什么？

### 人工智能在金融领域

人工智能本身是一个包罗万象的术语，它包含了许多技术进步，包括：

机器学习——使用神经网络、统计学和运筹学来识别数据中的洞察力，而不需要编程来得出结论。

深度学习——使用多层计算能力和改进的训练技术来识别数据中的模式。

在交付人工智能解决方案时，您可能会结合许多技术。例如，通过使用结构化和非结构化数据来增强分析和预测，从而提供前瞻性的思维洞察力。基于云的存储对于所涉及的数据量和所需的计算能力来说可能是一种资产。

各机构都有使用人工智能解决 RPA 过程中数据验证错误的例子，一些组织正在创建流程链，通过聊天机器人获取数据，使用 RPA 工具输入数据，并使用机器学习解决错误。

### 人工智能在金融中的应用

将人工智能应用到金融领域，对于许多组织来说是一个机会。这可能只是一个概念的证明，而不是一个可靠的解决方案。

对金融而言，或许最重要的问题是培养下一波技术浪潮所需的技能。人工智能需要理解数据、处理数据以及解决问题。这将把金融所需的技能提升到另一个水平。在提出解决方案之前，需要确定和纠正的一点是，当输入的数据集不能代表输出结果，机器以错误的方式"学习"时，会产生输出偏差。

# Chapter *10*

# Artificial Intelligence in Healthcare

## Text A

Every second *post-millennial* believes that they will work together with robots and artificial intelligence (AI) within 10 years. We *examine* the questions of what this desire means for the future of the *workforce*, and whether it has any *implications* for the Healthcare industry.

The Healthcare workforce crisis is due to at least three major issues: doctor shortages worldwide, the aging and *burnout* of physicians and a higher demand for chronic care. An effective system depends on the availability, accessibility, acceptability and quality of its health workers. It is estimated that the needs-based shortage of *Healthcare* workers *globally* is about 17.4 million, and the aging workforce is an additional challenge.

With the increase of life *expectancy* (the population over 65 years is expected to double by 2030) and the number of chronic illnesses, the demand towards the Healthcare system is also constantly growing. Consequently, the lack of access to care and the differing quality are general worldwide. 400 million people lack access to one or more *essential* health services, and five billion people do not have access to safe, *affordable surgical and anaesthesia* care when needed.

One in three physicians are over 55 years of age, and a third of physicians are expected to retire in the next decade. As the new generation of medical professionals looks for a limited number of

### New Words and Expressions

**post-millennial**
后千禧时代
**examine**/ɪɡˈzæmɪn; eg-/
检查；调查
**workforce**/ˈwɜːkfɔːs/
劳动力；工人总数
**implication**/ˌɪmplɪˈkeɪʃ(ə)n/
牵连；影响
**burnout**/ˈbɜːnaʊt/
烧坏；燃料烧尽
**Healthcare workers**
卫生工作者
**globally**/ˈɡləʊbəlɪ/
全球地；全局地
**expectancy**/ɪkˈspekt(ə)nsɪ ek-/
期望，期待
**essential**/ɪˈsenʃ(ə)l/
基本的；必要的
**affordable**/əˈfɔːdəbəl/
负担得起的
**surgical**/ˈsɜːdʒɪk(ə)l/
外科的；手术上的
**anaesthesia**/ˌænɪsˈθiːzɪə/
麻醉；麻木
**surgical and anaesthesia**
外科和麻醉

working hours, a speciality without having to be on-call and an acceptable work-life balance, this wish for a more controllable lifestyle might further increase shortages.

Because of the growing number of *chronic* patients and physicians being overloaded due to shortages, burnout is increasing. It can be the cause of somatic symptoms, substance use, psychological and sleep disorders; as well as *maladaptive* coping strategies. As physicians' *well being* is linked to the quality and the safety of outcomes, this further *fuels* challenges.

The human resource crisis is widening across the globe, and it is obvious that without a capable workforce there is no way to provide quality care. How can disruptive technologies in Healthcare help solve the variety of human resource problems? Will technology empower physicians or replace them? How can the medical curriculum, including post-graduate education prepare professionals for the meaningful use of technology?

These gaps have been growing for decades, and the promise of technology filling them is imminent with digital health becoming widespread. Authors of this essay argue that AI might not only fill the human resource gaps, but also raise ethical questions we need to assess today.

As Nick Bostrom describes in his book *super-intelligence*, AI is divided broadly into three stages: artificial *narrow* intelligence (ANI), artificial general intelligence (AGI) and artificial super-intelligence (ASI). In the next decade, ANI has the highest chance of being used in the medical practice for analyzing large datasets, finding new *correlations* and generally supporting *caregivers*' jobs.

An obvious first step is clearing up the definitions around AI to stop its misuse in medical *communication*. Here we attempt at providing short definitions for the most common expressions.

## Artificial Narrow Intelligence

It is good at performing a single task, such as playing chess, poker or Go, making purchase suggestions, online *searches*, sales predictions and *weather forecasts*.

## Artificial General Intelligence

It can understand and reason its environment similarly as a human being would do, therefore it's also known as human-level AI.

---

### New Words and Expressions

**chronic** /ˈkrɒnɪk/
慢性的；长期的；习惯性的

**maladaptive** /ˌmæləˈdæptɪv/
适应不良的；不适应的

**well being**
幸福；健康

**fuel** /fjʊəl/
燃料

**super-intelligence**
超级智能

**narrow** /ˈnærəʊ/
狭窄的，有限的

**correlation** /ˌkɒrəˈleɪʃ(ə)n/
相关；关联

**caregiver** /ˈkeəɡɪvə(r)/
照料者，护理者

**communication** /kəmjuːnɪˈkeɪʃ(ə)n/
通信

**search** /sɜːtʃ/
搜寻；调查；探求

**weather forecast**
天气预测，天气预报

## Artificial Super-intelligence

According to Nick Bostrom, it's smarter than the best humans in every field from scientific *creativity* to general *wisdom* and social skills.

## Supercomputers

A supercomputer is a computer with a high level of computing performance used for *resource-intensive* tasks, such as machine and deep learning.

## Machine Learning

Machine learning is one of the many subsets of AI that refers to creating programs based on data as opposed to programming rules. A software that learns from large sets of *relevant data* (e.g. feeding it with a lot of *radiology* images and letting it discover recurring patterns).

## Deep Learning

It is a specialized subset of machine learning that uses *neural* networks, an artificial replication of the *structure* and *functionality* of the brain. It's efficient at various tasks such as image recognition, natural language processing and translation. The performance of deep learning *algorithms* continues to improve as datasets grow *significantly* which means the bigger the dataset, the better the outcome and efficiency improves.

Various companies and organizations have already *demonstrated* how AI can *contribute to* improve the quality of care and/or decreasing costs.

Deepmind Health *launched* a cooperation with the Moorfields Eye Hospital NHS Foundation Trust to improve eye treatment by mining one million anonymized eye scans with the related medical records. IBM launched Watson Oncology to provide *clinicians* with *evidence-based* treatment options and an advanced ability to analyze the meaning and context of structured and unstructured data in clinical notes and reports.

In the Netherlands, Zorgprisma Publiek helps caregivers and hospitals avoid *unnecessary* hospitalizations of patients by analyzing the digital *invoices obtained* from *insurance* companies with IBM Watson in the cloud.

In *radiology*, the Medical Sieve project aims at building the *next-generation* "cognitive assistant" with analytical, reasoning

---

### New Words and Expressions

**creativity**/ˌkriːeɪˈtɪvɪtɪ/
创造力；创造性
**wisdom**/ˈwɪzdəm/
智慧，才智
**resource-intensive**
【环境】资源密集
**relevant data**
有关数据
**radiology**/reɪdɪˈɒlədʒɪ/
放射学；放射线科
**neural**/ˈnjʊər(ə)l/
神经的；神经系统的
**structure**/ˈstrʌktʃə/
结构；构造
**functionality**/fʌŋkʃəˈnælətɪ/
功能
**algorithms**/ˈælgərɪð(ə)mz/
算法；算法式
**significantly**/sɪɡˈnɪfɪk(ə)ntlɪ/
显著地；相当数量地
**demonstrate**/ˈdemənstreɪt/
论证；证明
**contribute to**
有助于；捐献
**launch**/lɔːntʃ/
发射；发行；开始
**clinician**/klɪˈnɪʃn/
临床医生
**evidence-based**
基于证据的
**unnecessary**/ʌnˈnesəs(ə)rɪ/
不必要的；多余的
**invoice**/ˈɪnvɒɪs/
发票；货物；发货单
**obtain**/əbˈteɪn/
获得；流行
**insurance**/ɪnˈʃʊər(ə)ns/
保险；保险费；保险契约；赔偿金
**radiology**/reɪdɪˈɒlədʒɪ/
放射学；放射线科
**next-generation**
下一代

capabilities and a range of clinical knowledge. Such an assistant would be able to analyze radiology images to detect medical issues. In *genomics*, Deep Genomics helps identify *linkages* to diseases in large data sets of genetic information and medical records.

In *pharmaceutical* research, Atomwise uses supercomputers to find new *therapies* speeding up clinical trials that take sometimes more than a decade and cost billions of dollars. As an example, Atomwise found two drugs predicted by the company's AI technology which may significantly reduce Ebola *infectivity* in less than a day of research, instead of years.

Deep learning algorithms have demonstrated to be able to help the *diagnosis* of conditions in *cardiology*, *dermatology* and *oncology*.

Arterys already received FDA clearance for its AI-assisted cardiac imaging system in 2017. AI supported messaging apps and voice controlled chatbots can also help take off the burden on medical professionals regarding easily diagnosable health concerns or quickly solvable health management issues. Safe drug bot is a chat messaging service that offers assistant-like support to health professionals who need appropriate information about the use of drugs during breastfeeding.

AI-based services could facilitate more accurate diagnoses, administration, decision-making, big data analytics, post-graduate education, among others. However, we need to emphasize that practicing medicine is not a linear process. Every single element and parameter cannot be translated into a programming language. Moreover, there is no *clinical trial* or peer-reviewed data about the data points that contribute to a medical decision. It's clear that AI is not the ultimate solution for all the challenges Healthcare faces today. Although, in many areas, its use is inevitable and advantageous in supporting caregivers' job.

However, a tight framework from regulatory agencies would further stop companies from providing false hope for patients claiming more than what they can *deliver* and prove. Moreover, the FDA has *assembled* a team of computer scientists and engineers to help oversee and anticipate future developments in AI-driven medical software. These are encouraging steps forward, but the range of *ethical*, legal and social *implications* of using AI in Healthcare are even beyond the *scope* of what we can deal with today.

---

### New Words and Expressions

**genomics**/dʒəˈnɒmɪks/
基因组学；基因体学

**linkage**/ˈlɪŋkɪdʒ/
连接；结合；联接；联动装置

**pharmaceutical** /ˌfɑːməˈsuːtɪk(ə)l/
制药的

**therapy**/ˈθerəpɪ/
治疗，疗法

**infectivity**/ˌinfekˈtivəti/
传染性（影响别人）；易传染

**clinical trial**
临床试验；诊治试验

**diagnosis**/ˌdaɪəgˈnəʊsɪs/
诊断

**cardiology**/kɑːdɪˈɒlədʒɪ/
心脏病学

**dermatology**/ˌdɜːməˈtɒlədʒɪ/
皮肤医学

**oncology**/ɒŋˈkɒlədʒɪ/
肿瘤学

**deliver**/dɪˈlɪvə/
交付；发表；递送

**assemble**/əˈsemb(ə)l/
集合，聚集

**ethical**/ˈeθɪk(ə)l/
伦理的；道德的；凭处方出售的；处方药

**implication**/ɪmplɪˈkeɪʃ(ə)n/
含义；暗示；牵连

**scope**/skəʊp/
范围；余地

Resource-poor regions will face challenges while adopting AI. On one hand, the cost of *disruptive* technologies might be too high for underdeveloped countries, pushing them further behind in improving Healthcare. This still stands if we consider that the use of new technologies could be *cost-effective* in the long run. If a country invests into buying an AI-based decision-support system, it could help physicians make better decisions, thus leading to fewer number of unnecessary hospitalizations, which reduces costs.

On the other hand, underdeveloped countries can be more open to policy changes that would *facilitate* the adoption of such technologies, which could lead to a more widespread adoption than in developed regions. Examples include how Rwanda opened up its emergency care system to Zipline that produces and operates medical *drones* across the country.

Regarding caregivers in general, do we need narrow or general AI to provide better care? What *elements* of their repetitive tasks, such as note taking or *administrative* duties could AI ease, and which ones such as diagnosis, treatment or monitoring could it facilitate?

Technology could also offer solutions to improve the access to care. With AI, it is easier for medical professionals to care for a larger number of patients. AI tools help them make better diagnostic decisions, improve treatment outcomes and reduce medical errors. AI could also take part in solving HR issues, such as *recruiting* and selecting the potential Healthcare workforce. It's important to point out that the HR crisis cannot be solved by only developing technologies for physicians. All Healthcare professionals must be involved.

However, AI does not cover the whole process of treatment: *empathy*, proper communication and the human touch are still equally essential. No application, software or device can replace personal connection and trust. The role of the human physician is inevitable, but AI could be a very useful cognitive assistant.

AI also means a paradigm shift in the doctor-patient relationship. As digital health transforms the well-known doctor-patient hierarchy into an equal level partnership, what happens with the autonomy that has been the essence of care? Who is responsible if an AI-assisted medical decision causes harm to a patient? Most doctors

*New Words and Expressions*

**disruptive**/dɪsˈrʌptɪv/
破坏的；分裂性的

**cost-effective**/ˈkɔːstəˈfektɪv/
划算的；成本效益好的

**facilitate**/fəˈsɪlɪteɪt/
促进；帮助；使容易

**drone**/drəʊn/
嗡嗡的声音

**elements**/ˈeləmənts/
基础；原理

**repetitive**/rɪˈpetɪtɪv/
重复的

**administrative**/ədˈmɪnɪstrətɪv/
管理的，行政的

**recruiting**/rɪˈkruːtɪŋ/
招聘；招募

**empathy**/ˈempəθɪ/
神入；移情作用

use online tools to help with research. Is there really a difference when it comes to using AI? Should AI be handled as another tool, such as a *stethoscope* or as an individual *entity*?

On the patients' side, will they stick to the human touch when shortages simply do not give them a chance to meet a *physician* in person for every medical issue? What if AI algorithms can *mimic* empathy either through an app or a chatbot? It's not yet known whether they will accept the use of AI in *decision-making* and learn its use in their care.

On the level of society, will it help shift focus from treatment to prevention? Will AI increase the cost of care? Will doctors and medical professionals be more efficient, because AI handles some of the *time-consuming* tasks? Will doctors provide better care in underdeveloped regions with the use of AI? And generally, how will it (if at all) change the current structures of insurance policies?

If we might be brave enough to *articulate* a vision, the authors of this debate article think that AI will eventually be evidence-based, widespread and affordable. Physicians have been translating the data they measured with rudimentary tools, like stethoscopes or blood pressure cuffs, and they will keep on doing the same with digital tattoo-like sensors and AI. We think this technology will reduce costs in providing care, making it faster and more efficient leading to a change in the medical profession that will involve more tasks related to creativity and critical thinking than time-consuming repetitions.

In about 20 years, 50% of jobs will be outdated or not needed anymore, and Healthcare is not an exception. While AI demonstrates significant potentials in improving diagnostics, it will probably not solve the HR crisis in Healthcare, or at least it will not start with that. The chance for improving the job environment and conditions of physicians is higher, which can eventually lead to a general improvement in the quality of care. If it becomes able to take over important tasks from medical professionals, it might even bring forward a *renaissance* era in the doctor-patient relationship.

While there are even more questions to address, our stand is that AI is not meant to replace medical professionals, but the ones using AI will probably replace those who don't. We also think that it is every caregivers' *duty* to prepare for a future like that.

---

### New Words and Expressions

**stethoscope**/ˈsteθəskəʊp/
听诊器

**entity**/ˈentɪtɪ/
实体；存在；本质

**physician**/fɪˈzɪʃ(ə)n/
医师；内科医师

**mimic**/ˈmɪmɪk/
模仿，摹拟

**decision-making**/dɪˈsɪʒən mekɪŋ/
决策的

**time-consuming**
/ˈtaimkənˌsju:mɪŋ/
耗时的；旷日持久的

**articulate**/ɑːˈtɪkjʊleɪt/
用关节连接；使相互连贯

**renaissance**/rɪˈneɪsns/
新生；再生；复活

**duty**/ˈdjuːtɪ/
责任；关税；职务

## Terms

### 1. Healthcare

Healthcare 是医疗保健是指通过预防、诊断和治疗人类的疾病、疾病、损伤和其他身心障碍来维持或改善健康。卫生保健由联合卫生领域的卫生专业人员（提供者或从业人员）提供。医生和医生助理是这些卫生专业人员的一部分。牙科、助产士、护理、医学、验光、听力学、药剂学、心理学、职业治疗、物理治疗和其他保健专业都是保健的一部分。它包括在提供初级保健、二级保健和三级保健以及公共卫生方面所做的工作。

### 2. Artificial Narrow Intelligence

人工狭义智能（弱人工智能），擅长完成特定任务，例如下棋、扑克或围棋，提出购买建议，在线搜索，销售预测和天气预报等。

### 3. Artificial General Intelligence

人工广义智能（强人工智能），能像人类一样理解和推理，因此也被称为人类水平的人工智能。

### 4. Artificial Superintelligence

人工超智能（超人工智能），根据尼克·博斯特罗姆（Nick Bostrom）的说法，从科学创造力到一般智慧和社交技能，在各个领域，它都比最优秀的人类更聪明。

### 5. Supercomputers

Supercomputers 是用于资源密集型任务(如机器和深度学习)的具有高计算性能的超级计算机。

### 6. Machine Learning

机器学习是人工智能的众多子集之一，它指的是基于数据而不是编程规则创建程序。从大量相关数据中学习的软件。例如：向它提供大量放射学图像，并让它发现重复出现的模式。

### 7. Deep Learning

Deep Learning 是机器学习的一个特殊子集，使用神经网络，人工复制大脑的结构和功能。它能有效地完成图像识别、自然语言处理和翻译等任务。随着数据集的显著增长，深度学习算法的性能不断提高，这意味着数据集越大，结果越好，效率越高。

## Comprehension

### Blank Filling

1. The healthcare workforce crisis is due to at least three major issues: _____ for chronic care.
2. An effective system depends on the _____ of its health workers.
3. _____ is good at performing a single task, such as playing chess, poker or Go, making purchase suggestions, online searches, sales predictions and weather forecasts.
4. A supercomputer is a computer with a high level of _____ used for resource-intensive tasks, such as machine and deep learning.

5. _____ is one of the many subsets of AI that refers to creating programs based on data as opposed to programming rules.

6. _____ can understand and reason its environment similarly as a human being would do, therefore it's also known as human-level AI.

**Content Questions**

1. AI-based services could facilitate more accurate diagnoses, administration, decision-making, big data analytics, post-graduate education, among others. However, we need to?

## Answers

**Blank Filling**

1. doctor shortages worldwide, the aging and burnout of physicians and a higher demand

2. availability, accessibility, acceptability and quality

3. computing performance

4. computing performance

5. Machine learning

6. Artificial general intelligence

**Content Questions**

1. We need to emphasize that practicing medicine is not a linear process. Every single element and parameter cannot be translated into a programming language. Moreover, there is no clinical trial or peer-reviewed data about the data points that contribute to a medical decision. It's clear that AI is not the ultimate solution for all the challenges healthcare faces today. Although, in many areas, its use is inevitable and advantageous in supporting caregivers' job.

## 参考译文 A

"后千禧一代"一直相信，他们将在十年内与机器人和人工智能（AI）合作。我们将面临的问题是，这种愿景对未来的劳动力意味着什么，以及它是否对医疗行业有影响。

卫生保健工作人员危机至少是由三个主要问题造成的：世界范围内的医生短缺、医生的老龄化和职业倦怠以及对慢性病护理的更高需求。一个有效的系统取决于其卫生工作者的可用性、可及性、可接受性和质量。据估计，全球以需求为基础的医疗工作者短缺约 1740 万人，而老龄化的劳动力是第三个挑战。

随着预期寿命的增长（到 2030 年，65 岁以上的人口预计将翻一番）和慢性病的数量，对医疗保健系统的需求也在不断增长。因此，缺乏获得护理的机会和不同的质量在世界范围内是普遍的。4 亿人无法获得一项或多项基本卫生服务，50 亿人在需要时无法获得安全、负担得起的手术和麻醉护理。

三分之一的医生年龄在 55 岁以上，三分之一的医生预计在未来十年退休。随着新一代医学专业人士寻求有限的工作时间、无须随叫随到的专业以及可接受的工作与生活平衡，这种对更可控生活方式的渴望可能会进一步加剧短缺。

由于越来越多的慢性病患者和医生由于人手不足而超负荷工作，他们的工作倦怠也在增加。它可引起躯体症状、物质使用、心理和睡眠障碍；以及适应不良的应对策略。由于医生的健康与结果的质量和安全性有关，这进一步加剧了挑战。

人力资源危机正在全球范围内蔓延，很明显，没有一支有能力的劳动力队伍，就无法提供高质量的医疗服务。医疗保健中的颠覆性技术如何帮助解决各种各样的人力资源问题？技术是赋予医生权力还是取代他们？包括研究生教育在内的医学课程如何为专业人员有意义地使用技术做好准备？

几十年来，这些差距一直在扩大，随着数字健康的普及，技术填补这些差距的希望迫在眉睫。这篇文章的作者认为人工智能可能不仅填补了人力资源的缺口，而且提出了我们今天需要评估的伦理问题。

正如尼克·博斯特罗姆（Nick Bostrom）在他的《超智能》（*Superintelligence*）一书中所描述的，人工智能大致分为三个阶段：人工狭义智能（ANI）、人工通用智能（AGI）和人工超智能（ASI）。在接下来的十年里，ANI 最有可能在医疗实践中被用于分析大型数据集，发现新的相关性，并普遍支持护理人员的工作。

显然第一步是澄清人工智能的定义，以阻止它在医学交流中的滥用。在这里，我们试图为最常见的表达式提供简短的定义。

### 人工狭义智能

它擅长完成单个任务，例如下棋、扑克或围棋，提出购买建议，在线搜索，销售预测和天气预报。

### 人工通用智能

它能像人类一样理解和推理环境，因此也被称为人类水平的人工智能。

### 人工超级智能

根据尼克·博斯特罗姆（Nick Bostrom）的说法，从科学创造力到一般智慧和社交技能，在各个领域，它都比最优秀的人类更聪明。

### 超级计算机

超级计算机是用于资源密集型任务（如机器和深度学习）的具有高计算性能的计算机。

### 机器学习

机器学习是人工智能的众多子集之一，它指的是基于数据而不是编程规则创建程序。从大量相关数据中学习的软件（例如，向它提供大量放射学图像，并让它发现重复出现的模式）。

### 深度学习

它是机器学习的一个特殊子集，使用神经网络，人工复制大脑的结构和功能。它能有效地完成图像识别、自然语言处理和翻译等任务。随着数据集的显著增长，深度学习算法的性能不断提高，这意味着数据集越大，结果越好，效率越高。

许多公司和组织已经展示了人工智能如何有助于提高护理质量和/或降低成本。

**Deepmind Health** 与 **Moorfields** 眼科医院 **NHS** 基金会信托合作，通过挖掘一百万份匿名的眼部扫描和相关的医疗记录来改善眼科治疗。**IBM** 推出 **Watson** 肿瘤学，为临床医生提供基于证据的治疗选择，并提供先进的能力来分析临床笔记和报告中结构化和非结构化数据的含义。

在荷兰，Zorgprisma Publiek 通过分析保险公司在云端使用 IBM Watson 获得的数字证书，帮助护理人员和医院避免不必要的病人住院。

在放射学方面，医学筛选项目的目标是构建具有分析、推理能力和一系列临床知识的下一代"认知助手"。这样的助手将能够分析放射学图像，以发现医学问题。在基因组学中，深度基因组学有助于在遗传信息和医疗记录的大数据集中识别与疾病的联系。

在药物研究中，Atomwise 利用超级计算机寻找新疗法，加快临床试验的速度，有时需要 10 年以上的时间，耗资数十亿美元。例如，Atomwise 公司发现了该公司人工智能技术预测的两种药物，这两种药物可能在不到一天的时间内（而不是几年的时间），显著降低埃博拉病毒的传染性。

深度学习算法已经证明能够帮助诊断心脏病、皮肤病和肿瘤学。

Arterys 公司已经在 2017 年获得了 FDA 对其人工智能辅助心脏成像系统的许可。人工智能支持的消息应用程序和语音控制聊天机器人也可以帮助减轻医疗专业人士在容易诊断的健康问题或快速解决的健康管理问题上的负担。Safedrugbot 是一款聊天信息服务，为需要了解哺乳期间药物使用情况的医护人员提供类似助手的支持。

基于人工智能的服务可以促进更准确的诊断、管理、决策、大数据分析、研究生教育等。然而，我们需要强调的是，行医不是一个线性的过程。不能将每个元素和参数都翻译成编程语言。此外，没有临床试验或同行评议的数据，这些数据点有助于作出医疗决定。很明显，人工智能并不是医疗保健面临的所有挑战的最终解决方案。虽然，在许多领域，它的使用是不可避免的，有利于支持护理人员的工作。

然而，来自监管机构的严格框架将进一步阻止企业为患者提供虚假的希望，这些患者声称他们所能提供和证明的东西超出了他们的能力。此外，FDA 还组建了一个由计算机科学家和工程师组成的团队，帮助监督和预测人工智能驱动的医疗软件的未来发展。这些都是令人鼓舞的进步，但在医疗保健中使用人工智能，这在伦理、法律和社会影响等方面，超出了我们今天能够控制的范围。

人工智能在资源贫乏地区的应用将面临挑战。一方面，颠覆性技术的成本对不发达国家来说可能过高，从而在改善医疗保健方面更不积极。如果我们认为新技术的使用从长期来看是具有成本效益的，这一点仍然成立。如果一个国家投资购买一个基于人工智能的决策支持系统，它可以帮助医生做出更好的决策，从而减少不必要的住院次数，从而降低成本。

另一方面，不发达国家可以更开放地接受有利于采用这种技术的政策变化，这可能导致比发达区域更广泛地采用这种技术。例如，卢旺达如何开放其紧急护理系统，以便在全国范围内生产和运营医用无人机。

对于普通的看护人，我们需要狭义的人工智能还是广义的人工智能来提供更好的协助？人工智能可以做哪些重复性工作？如做笔记或行政。哪些任务通过采用人工智能可以受益？如诊断，治疗或监测。

技术也可以提供解决方案，改善获得护理的机会。有了人工智能，医护人员更容易照顾更多的病人。人工智能工具帮助他们做出更好的诊断决定，改善治疗结果，减少医疗差错。人工智能还可以参与解决人力资源问题，例如招聘和选择潜在的医疗人力资源。需要指出的是，人力资源危机不能仅仅通过为医生开发技术来解决。所有医疗专业人员都必须参与其中。

然而，人工智能并没有涵盖整个治疗过程：同情心、适当地沟通和人际接触仍然同样重要。任何应用程序、软件或设备都不能取代人与人之间的联系和信任。人类医生的角色是不可避免的，但人工智能可能是一个非常有用的认知助手。

人工智能还意味着医患关系的范式转变。随着数字健康将著名的医患关系转变为平等的合作关系，作为护理本质的自主权会发生什么变化？如果人工智能辅助医疗决定对病人造成伤害，谁负责？大多数医生使用在线工具帮助研究。使用 AI 真的有区别吗？人工智能应该作为像听诊器一样的工具来处理，还是作为一个单独的实体来处理？

从患者的角度来看，如果药物短缺不能让他们有机会就每一个医疗问题与医生面谈，他们会坚持人性化的交流吗？如果人工智能算法可以通过应用程序或聊天机器人模拟同理心会怎样？目前还不清楚他们是否会接受人工智能在决策过程中的使用，以及学习人工智能在他们的护理中的使用。

在社会层面上，它是否有助于将重点从治疗转移到预防？人工智能会增加护理成本吗？因为人工智能处理一些耗时的任务，医生和医疗专业人士会更有效率吗？在欠发达地区，医生使用人工智能会提供更好的医疗服务吗？总的来说，它（如果有的话）将如何改变当前的保险结构？

如果我们有足够的勇气实现一个愿景，这篇辨文的作者认为，人工智能的实现最终是基于可验证性、普遍性和可负担性三个基本指标。医生们一直在用听诊器或血压袖带等基本工具转换他们测量的数据，他们还将继续用类似纹身的数字传感器和人工智能做同样的事情。我们认为，这项技术将降低提供医疗服务的成本，使之更快、更有效，从而改变医疗行业，这将涉及更多与创造力和批判性思维有关的任务，而不是耗时的重复工作。

大约 20 年后，50%的工作岗位将过时或不再需要，医疗保健也不例外。虽然人工智能在改善诊断方面显示出了巨大的潜力，但它可能无法解决医疗行业的人力资源危机，或者至少不会从解决危机开始。改善医生工作环境和条件的机会比较大，这最终会导致护理质量的普遍提高。如果它能够从医学专业人士手中接过重要的任务，甚至可能将医患关系带入一个复兴的时代。

虽然还有更多的问题需要解决，但我们的立场是，人工智能并不是要取代医学专业人士，但使用人工智能的人可能会取代那些不使用人工智能的人。我们还认为，每个护理人员都有责任为这样的未来做好准备。

# Text B

Recently AI techniques have sent vast waves across Healthcare, even *fuelling* an active discussion of whether AI doctors will eventually replace human physicians in the future. We believe that human physicians will not be replaced by machines in the *foreseeable* future, but AI can definitely assist physicians to make better clinical

*New Words and Expressions*

**fuelling**/ˈfjuəliŋ/

　v. 加油（fuel 的现在分词）；加燃料

　n. 油；燃料

**foreseeable**/fɔrˈsiəbl/ adj.

　可预知的；能预测的

dccisions or even replace human judgement in certain *functional areas* of Healthcare (e.g., radiology). The increasing availability of Healthcare data and rapid development of big data analytic methods has made possible the recent successful applications of AI in Healthcare. Guided by relevant clinical questions, powerful AI techniques can unlock *clinically* relevant information hidden in the *massive* amount of data, which in turn can assist clinical decision making.

In this article, we survey the current status of AI in Healthcare, as well as discuss its future. We first briefly review four relevant aspects from medical investigators' perspectives:

- motivations of applying AI in Healthcare
- data types that have be analysed by AI systems
- mechanisms that enable AI systems to generate clinical meaningful results
- disease types that the AI communities are currently *tackling*.

## Motivation

The advantages of AI have been extensively discussed in the medical literature. AI can use sophisticated algorithms to "learn" features from a large volume of Healthcare data, and then use the obtained insights to assist clinical practice. It can also be equipped with learning and self-correcting abilities to improve its accuracy based on feedback. An AI system can assist physicians by providing up-to-date medical information from journals, textbooks and clinical practices to inform proper patient care. In addition, an AI system can help to reduce diagnostic and therapeutic errors that are inevitable in the human clinical practice. Moreover, an AI system extracts useful information from a large patient population to assist making real-time inferences for health risk alert and health outcome prediction.

## Healthcare Data

Before AI systems can be deployed in Healthcare applications, they need to be "trained" through data that are generated from clinical activities, such as screening, diagnosis, treatment assignment and so on, so that they can learn similar groups of subjects, associations between subject features and outcomes of interest. These clinical data often exist in but not limited to the form of demographics, medical notes, *electronic* recordings from medical devices, physical examinations and clinical laboratory and images.

---

### New Words and Expressions

**functional area**
功能区；职能范围

**massive**/ˈmæsɪv/ adj.
大量的；巨大的；厚重的；魁伟的

**clinically**/ˈklinikli/ adj.
临床地；门诊部地

**tackling**/ˈtæklɪŋ/
处理

**electronic**/ɪˌlekˈtrɒnɪk/ adj.
电子的

Specifically, in the diagnosis stage, a substantial proportion of the AI literature analyses data from diagnosis imaging, genetic testing and *electrodiagnosis* (figure 1). For example, Jha and Topol urged radiologists to adopt AI technologies when analysing diagnostic images that contain vast data information. Li et al studied the uses of abnormal genetic expression in long non-coding RNAs to diagnose gastric cancer. Shin et al developed an electrodiagnosis support system for *localising* neural injury.

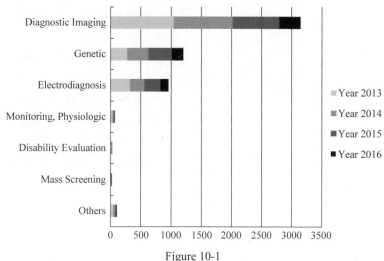

Figure 10-1

In addition, physical examination notes and clinical laboratory results are the other two major data sources (figure 1). We distinguish them with image, genetic and *ElectroPhysiological* (EP) data because they contain large portions of unstructured narrative texts, such as clinical notes, that are not directly analysable. As a consequence, the corresponding AI applications focus on first converting the unstructured text to machine-understandable electronic medical record (EMR). For example, Karakülah et al used AI technologies to extract *phenotypic* features from case reports to enhance the diagnosis accuracy of the congenital anomalies.

## AI Devices

The above discussion suggests that AI devices mainly fall into two major categories. The first category includes machine learning (ML) techniques that analyze structured data such as imaging, genetic and EP data. In the medical applications, the ML procedures attempt to *cluster* patients' traits, or infer the probability of the disease outcomes. The second category includes natural language

***New Words and Expressions***

**electrodiagnosis**
/elektrəʊdaɪəɡ'nəʊsɪs/ adj.
电反应诊断的；电诊法的
**localise**/'ləʊkəlaɪz/ vt.
使地方化；定位；集中
**electrophysiological**
/iˈlektrəʊˌfɪziəlɒdʒikəl/ adj.
电生理学的
**corresponding**/ˌkɒrɪ'spɒndɪŋ/ adj.
相当的，相应的；一致的；通信的
**phenotypic**/ˌfinə'tɪpɪk/ adj.
表型的
**cluster**/'klʌstə/ v.
群聚；丛生

processing (NLP) methods that extract information from unstructured data such as clinical notes/ medical journals to *supplement* and enrich structured medical data. The NLP procedures target at turning texts to *machine-readable structured* data, which can then be analysed by ML techniques.

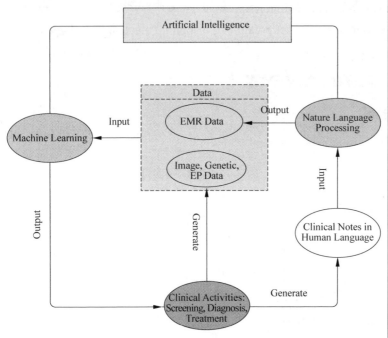

Figure 10-2

***New Words and Expressions***

**supplement**/ˈsʌplɪm(ə)nt/
增补，补充

**machine-readable**
/məˈʃiːnˈriːdəbl/ adj.
机器可读的；可用计算机处理的

**structure**/ˈstrʌktʃə/ n.
结构；构造；建筑物

**data generation**
数据生成

**enrichment**/ɪnˈrɪtʃmənt/ n.
丰富；改进；肥沃；发财致富

**cardiovascular**
/ˌkɑːdɪəʊˈvæskjʊlə/ adj.
心血管的

**doubleblinded**/ˈdʌblˌblaɪnd/ adj.
双盲

For better presentation, the flow chart in figure 2 describes the road map from clinical *data generation*, through NLP data *enrichment* and ML data analysis, to clinical decision making. We comment that the road map starts and ends with clinical activities. As powerful as AI techniques can be, they have to be motivated by clinical problems and be applied to assist clinical practice in the end.

**Disease Focus**

Despite the increasingly rich AI literature in Healthcare, the research mainly concentrates around a few disease types: cancer, nervous system disease and *cardiovascular* disease (figure 3). We discuss several examples below.

**1. Cancer**

Somashekhar et al demonstrated that the IBM Watson for oncology would be a reliable AI system for assisting the diagnosis of cancer through a *doubleblinded* validation study. Esteva et al analysed clinical images to identify skin cancer subtypes.

## 2. Neurology

Bouton et al developed an AI system to restore the control of movement in patients with quadriplegia. Farina et al tested the power of an offline man/machine interface that uses the discharge timings of spinal motor neurons to control upper-limb *prostheses*.

## 3. Cardiology

Dilsizian and Siegel discussed the potential application of the AI system to diagnose the heart disease through *cardiac* image. Arterys recently received clearance from the US Food and Drug Administration (FDA) to market its Arterys Cardio DL application, which uses AI to provide automated, editable ventricle *segmentations* based on conventional cardiac MRI images.

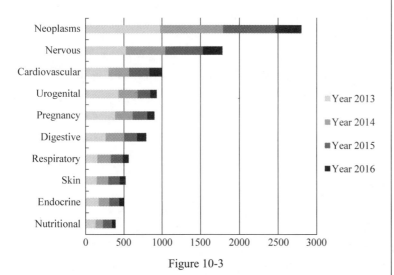

Figure 10-3

The *concentration* around these three diseases is not completely unexpected. All three diseases are leading causes of death; therefore, early diagnoses are crucial to prevent the deterioration of patients' health status. Furthermore, early diagnoses can be potentially achieved through improving the analysis procedures on imaging, genetic, EP or EMR, which is the strength of the AI system. Besides the three major diseases, AI has been applied in other diseases as well. Two very recent examples were Long et al, who analysed the ocular image data to diagnose *congenital* cataract disease, and Gulshan et al, who detected referable *diabetic retinopathy* through the retinal fundus photographs.

---

**New Words and Expressions**

**prostheses**/prɒsˈθiːsɪs/ n.
假体

**cardiac**/ˈkɑːdɪæk/
强心剂；心脏的；心脏病的

**segmentation**/ˌsegmənˈteɪʃən/
分割；割断；细胞分裂

**concentration**
/kɒns(ə)nˈtreɪʃ(ə)n/ n.
浓度；集中；浓缩；专心；集合

**congenital**/kənˈdʒenɪt(ə)l/
先天的，天生的；天赋的

**diabetic**/daɪəˈbetɪk/ adj.
糖尿病的；患糖尿病的

**retinopathy**/ˌretinˈɒpəθi, -ˈnɒp-/ n.
视网膜病

## 参考译文 B

最近，人工智能技术在医疗保健领域掀起了一股热潮，甚至引发了一场关于人工智能医生是否最终将取代人类医生的热烈讨论。我们相信在可预见的未来，人类医生不会被机器取代，但是人工智能绝对可以帮助医生做出更好的临床决策，甚至在医疗保健的某些功能领域（如放射学），人工智能可以代替人类的判断。医疗数据的日益普及和大数据分析方法的快速发展，使得人工智能在医疗领域的成功应用成为可能。在相关临床问题的指导下，强大的人工智能技术可以解开隐藏在海量数据中的临床相关信息，进而辅助临床决策。

本文综述了人工智能在医疗卫生领域的现状，并对人工智能的发展前景进行了展望。我们首先从医学研究者的角度简要回顾了四个相关方面：

- 人工智能在医疗中的应用动机；
- 人工智能系统分析的数据类型；
- 使人工智能系统产生有意义的临床结果的机制；
- AI 社区目前正在应对的疾病类型。

### 动机

人工智能的优点已在医学文献中广泛讨论。人工智能可以使用复杂的算法从大量医疗数据中"学习"特征，然后利用获得的洞见帮助临床实践。还可以通过学习和自我纠错来提高基于反馈的准确性。人工智能系统可以通过提供来自期刊、教科书和临床实践的最新医疗信息来帮助医生提供适当的病人护理。此外，人工智能系统可以帮助减少在人类临床实践中不可避免的诊断和治疗错误。此外，人工智能系统从大量患者中提取有用信息，帮助实时推断健康风险预警和健康结果预测。

### 医疗数据

在人工智能系统被部署到医疗保健应用程序之前，它们需要通过临床活动（如筛选、诊断、治疗分配等）生成的数据进行"培训"，以便它们能够学习相似的主题组、主题特征与感兴趣的结果之间的关联。这些临床数据主要基于以下（但不限于）形式存在：人口统计、病历、医疗器械电子记录、体检、临床实验室和影像等。

具体来说，在诊断阶段，相当一部分 AI 分析来自诊断成像、基因检测和电诊断的数据（图 10-1），例如 Jha 和 Topol 在分析包含大量数据信息的诊断图像时，敦促放射科医师采用 AI 技术。Li 等研究了长链非编码 RNA 中异常基因表达在胃癌诊断中的应用。Shin 等开发了一种定位神经损伤的电诊断支持系统。

为了更好地表达，图 10-2 中的流程图描述了从临床数据生成，通过 NLP 数据丰富和 ML 数据分析，到临床决策的路线图。我们说，路线图从临床活动开始和结束。人工智能技术虽然强大，但必须以临床问题为动力，并最终应用于辅助临床实践。

### 疾病的关注

尽管人工智能在医疗领域的文献越来越丰富，但研究主要集中在几种疾病类型上，如癌症、神经系统疾病和心血管疾病（图 10-3）。

### 1. 癌症

Somashekhar 等证明 IBM 肿瘤沃森将是一个可靠的人工智能系统，通过双盲验证研究来辅助癌症诊断。Esteva 等分析了临床图像以识别皮肤癌亚型。

### 2. 神经病学

薄敦等开发了一种人工智能系统，用于恢复四肢瘫痪患者的运动控制。Farina 等测试了一种利用脊髓运动神经元放电计时来控制上肢假肢的人机界面的功率。

### 3. 心脏病学

Dilsizian 和 Siegel 讨论了人工智能系统通过心脏图像诊断心脏病的潜在应用。Arterys 公司最近获得了美国食品和药物管理局（FDA）的许可，可以销售其 Arterys 心脏 DL 应用程序，该应用程序使用 AI 提供基于常规心脏 MRI 图像的自动化、可编辑的心室分割。

人们理解为什么研究集中于这三种疾病。这三种疾病都是导致死亡的主要原因；因此，早期诊断是防止患者健康状况恶化的关键。此外，通过改进图像、遗传、EP 或 EMR 的分析程序，可以实现早期诊断，这是人工智能系统的强项。除了这三种主要疾病，人工智能还被应用于其他疾病。最近的两个例子是 Long 等人，他们通过分析眼部图像数据来诊断先天性白内障，Gulshan 等人通过视网膜眼底照片发现了可参考的糖尿病视网膜病变。

# Chapter *11*

# Artificial Intelligence in Transportation

## Text A

Faced with dwindling *fossil* fuels, and the increasingly negative impact of climate change on society, several countries have instigated national plans to reduce carbon *emissions*. In particular, the electrification of transport is seen as one of the main pathways to achieve significant reductions in $CO_2$ emissions. In the last few years EVs. have gained ground, and, to date, more than 180 thousand of them have been deployed worldwide. Despite this number corresponding to only 0. 02% of all *vehicles* on the roads, an ambitious target of having over 20 million EVs. on the roads by 2020 has been set by the International Energy Agency.

In order to ensure that the largescale deployment of EVs (results in a significant reduction of $CO_2$ emissions) it is important that they are charged using energy from *renewable sources* (e.g., wind, solar). Crucially, given the *intermittency* of these sources, mechanisms, as part of a Smart *Grid*, need to be developed to ensure the smooth integration of such sources in our energy systems. EVs. could potentially help by storing energy when there is a surplus, and feed this energy back to the grid when there is demand for it.

Indeed, the ability of EVs. to store energy while being used for transportation represents an enormous potential to make energy systems more efficient. On the one hand, given that *vehicles* drive only for a small percentage of the day (4-5% in the US), and a large percentage of the vehicles stay unused in parking lots (90% in the

**New Words and Expressions**

**fossil**/ˈfɒs(ə)l; -sɪl/ adj.

僵化的事物；化石的

**emission**/ɪˈmɪʃən/ n.

排放，辐射

**vehicle**/viːɪkəl/ n.

交通工具

**renewable sources**

再生能源

**intermittency**

/ˌɪntəˈmɪtəns,-tənsi/ n.

间歇性；间歇现象

**grid**/grɪd/ n.

网格；格子

**vehicles**/ˈviːɪklz/ n.

运载工具

US), and considering the fact that EVs. are equipped with large batteries, they could be used as storage devices when parked (e.g., as part of Vehicle-to-Grid(2G) schemes), and thus *dramatically* increase the storage capacity of the network. Indeed, studies have shown that if one fourth of vehicles in the US were electric this would double the current storage capacity of the network. On the other hand, given that large numbers of EVs. need to charge on a daily basis, (40% of EV owners in California travel daily further than the range of their fully charged battery). EVs. charge as and when needed, they may overload the network. For this reason, new mechanisms are required to be able to manage the charging of EVs-Grid- to- Vehicle( G2V) - in real time while considering the constraints of the distribution networks within which EVs. need to charge. Moreover, EV routing systems should consider the ability of EVs. to recuperate energy while braking and/ or when driving downhill, and choose routes that fully utilise this ability. By so doing, it may be possible for EVs. to charge less often, thus maximising their range, reducing the costs for their owners, and minimising the peaks they cause on energy grids.

Against this background, a number of techniques and mechanisms to manage EVs, either individually or collectively, have been developed. For example, a number of web and mobile- based applications have been developed to provide information to EV drivers about the locations of charging points where available charging slots exist. Moreover, *prototype* systems for energy efficient routing have been developed while new types of chargers that can fully charge an EV battery in less than an hour are becoming commonplace. Thus, while a number of advances have been made in terms of the physical *infrastructure* and technologies for EVs., these may not be sufficient to manage the dynamism and uncertainty underlying the behaviour of individual and collectives of EVs. Controling the activities of EVs. will demand algorithms that can solve problems that involve a large number of *heterogeneous* entities (e.g., EV owners, charging point owners, grid operators), each one having its own goals, needs and incentives (e.g., amount of energy to charge, profit maximisation), while they will operate in highly dynamic environments (e.g., variable number of EVs. variable intentions of the drivers) and having to deal with a number of *uncertainties* (e.g.,

**New Words and Expressions**

**dramatically**/drəˈmætɪkəlɪ/ adv.
显著地，剧烈地

**prototype**/ˈprotəˈtaɪp/ n.
原型；标准，模范

**Infrastructure**/ˈɪnfrəˈstrʌktʃə/ n.
基础设施

**heterogeneous**
/ˌhet(ə)rə(ʊ)ˈdʒiːnɪəs; -ˈdʒen-/ adj.
不均匀的；由不同成分形成的

**uncertainties**/ʌnˈsətntɪ/ n.
不确定性；不确定因素

future arrival of EVs., future energy demand, energy production from renewable sources). Some of these challenges have recently been tackled by the Artificial Intelligence community, and in this paper we survey the state of the art of such AI approaches in the following EV application issues:

Energy efficient EV routing and range maximisation: algorithms and mechanisms have been developed to route EVs. in order to *minimise* energy loss and maximise energy harvested during a trip. In particular, building upon existing search algorithms, solutions have been developed to adapt to the needs and the characteristics of EVs., so as to take advantage of their energy recuperation ability and maximise the driving range. For example, and propose algorithms for energy efficient EV routing with or without recharging, while provides an algorithm for calculating reachable locations from a certain starting point given an initial battery level. Moreover, enhances the use of *super capacitors* with machine learning and data mining techniques to maximise the range of EVs.

Congestion management: algorithms have been designed to manage and control the charging of the EVs., so as to minimise queues at charging points, and the discomfort to the drivers. For example and propose algorithms for routing EVs. to charging points where the least congestion exists, considering the preferences and the constraints of the drivers (e.g., final destination, amount of electricity to charge) , while presents a *heuristic* algorithm to place charging points given a certain *topology* so that an EV is able to travel between any two locations without running out of energy.

Integrating EVs. into the Smart Grid: a number of mechanisms have been developed to schedule and control the charging of the EVs. (G2V) so that peaks and possible overloads of the electricity network may be avoided, while minimizing electricity cost. Moreover, we also survey approaches that utilise the storage capacity of the EVs. (V2G) in order to balance the electricity demand of various locations in the network, or to ease the integration of intermittent renewable energy sources to the grid.

In order to clarify the intersections and differences between the above challenges at a conceptual level, we provide an abstract description of the research *landscape* in Figure 1. While we use a tree representation (signifying a delineation between the concepts),

**New Words and Expressions**

**minimise**
轻描淡写

**super capacitors**
超级电容器

**heuristic**/ˌhjʊ(ə)ˈrɪstɪk/ adj.
启发式的；探索的

**topology**/təˈpɒlədʒɪ/ n.
拓扑学

**landscape**/ˈlændskeɪp/ n. vt.
风景；对…做景观美化；美化（环境等）

it is clear that there are overlaps (e.g., in terms of *congestion* management) between the different nodes of the tree.

*New Words and Expressions*

**congestion**/kənˈdʒestʃ(ə)n/ n.
拥挤；拥塞
**mechanism**/ˈmɛkənɪzəm/ n.
机制；原理，途径
**departure**/dɪˈpɑrtʃə/ n.
离开，离去

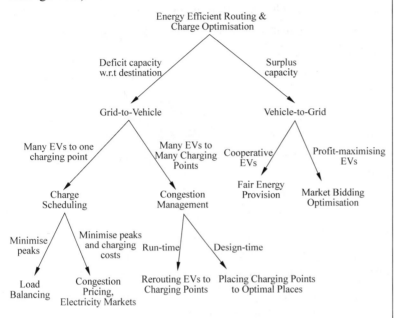

Figure 11-1　The electric vehicles research landscape

Thus, from this representation of the research landscape it can be seen that there are different considerations depending on whether the EVs. can travel or not based on their battery level (e.g., they need to route to their destination or charge) , which in turn, gives rise to challenges for Grid- to- vehicle and Vehicle-to-grid systems in terms of load balancing or congestion management among others. Coupled with such issues, is the problem of incentivising EV owners to take certain routes, charge at certain times (e.g., to avoid peaks) or to form part of EV collectives to trade on the energy markets. Finally, the infrastructure also needs to be designed in order to handle large numbers of EVs. (e.g., by placing charging points in appropriate places) whichever *mechanism* is used to charge EVs. or sell their spare capacity to the grid.

Against this background we can identify some key scientific dimensions of the problems that need to be tackled:

**1. Uncertainty**

While several algorithms have been proposed to account for uncertainty in renewable energy production, very few tackle the uncertainty in arrival and *departure* times of EVs. and the load they will impose on the distribution network, as well as uncertainties in the reliability of communication systems used to coordinate collectives of EVs.. The challenge here is to produce predictions at short notice as late decisions could potentially result in maior disruption to the transportation network. Hence. efficient machine learning *algorithms* need to be developed to predict behaviours in the system. In particular, we believe predictions could be improved by constructing better models of human mobility as well as by *fusing data* from across the transportation network. To this end, large scale deployments of EVs. are essential. should future machine learning algorithms be trained and evaluated in big enough datasets so as their efficiency to be maximised.

### 2. Dynamism

The state of the electricity grid, the production of renewable sources, the charging point availability, the congestion at communication and transportation networks and the number of EVs. available to provide V2G services, change quickly while a large number of EVs. are either driving or charging. Under such a dynamic setting, fail- safe mechanisms and approximation algorithms will be required to solve optimisation problems at short notice, while minimising communication bandwidth. While we have noted a number of solutions that use stochastic dynamic programming, it will be interesting to see if such solutions can be *decentralised* to ensure the system is more *robust*.

Furthermore, we can identify some key engineering dimensions of the problems that need to be tackled:

### 1. Interoperability

There is a need EV technologies to be able to work seamlessly and efficiently together. Different types of chargers should be able to work with all EV models, and data exchanged between entities EVs., charging points, network operators should have an understandable by all format and meaning. For example, Semantic Web technologies, such as XML, RDF and *ontologies*, can provide a structured and consistent way to represent the data being exchanged, and therefore make the collaboration of various technologies more efficient.

### 2. Privacy

In order for EVs. to be efficiently managed in terms of driving, charging and/or discharging, data on the location and the preferences of them must, in many cases, be obtained by a central mechanism. This creates issues of privacy and data protection, as drivers might not be willing to disclose such information.

---

*New Words and Expressions*

**algorithm**/ˈælgəˈrɪðəm/
运算法则

**fusing data**
融合的数据

**decentralise**/diːˈsentrəlais/
下放

**robust**/rəʊˈbʌst/ adj.
强健的

**ontologies**/ˈɒntɒlədʒɪs/ n.
本体论；知识本体；形而上学

### 3. Real world validation

Currently, most of the mechanisms and the technologies related to the management of the EVs. remain at a theoretical or at a pilot deployment level, and thus, their effectiveness in a large scale deployment has not been validated. The design of effective interfaces for human-ev (agent) interaction to be smooth and efficient, but also the research on ways to *motivate and incentivize* people to follow the instructions given to them by systems (e.g., a routing system giving instruction on an energy efficient route to take, or a charging point to charge at) is crucial. Moreover, the complexity of the coordination of a large number of entities (e.g., EVs., charging points, electricity network managers), and the ability of the systems to react to unexpected situations (e.g., a large number of EVs. wanting to charge within a short period of time) and prevent negative events (e.g., overloading of the electricity network) must be carefully studied, analyzed, and *verified*.

Our study has shown that several AI-based approaches are emerging in all areas of EV management: from battery charging *algorithms* to network congestion management algorithms. We believe in a concerted effort, involving transportation engineers, power systems experts, and AI researchers, in order to bring the benefits of such solutions to the real world. Hence, we advocate more joint deployments of novel AT solutions in field trials, with users of different types, in order to unpack more specific challenges that remain to be addressed before EVs. can be deployed at scale. Moreover, AI techniques being exploratory in nature can help EV researchers to quickly explore optimisation search spaces using *heuristics*. Since modern AI is being based on solid scientific approaches all its experiments and results are *verifiable* and *reproducible*. Thus, engineers can base the development of standards, which are crucial should a systematic management of EVs. activities be achieved, on the results of AI research on EVs. Currently, a number of EV related standards already exist. 20 and others are under way.

| New Words and Expressions |
| --- |
| **motivate and incentivize** |
| 激励和鼓励 |
| **verified**/ˈverɪfaɪd/ adj. |
| 已查清的，已证实的 |
| **algorithms**/ˈælgərɪð(ə)mz/ n. |
| 运算法则 |
| **heuristics**/hjuˈrɪstɪks/ n. |
| 启发（法），探索法 |
| **verifiable**/ˈvɛrɪfaɪəbl/ adj. |
| 能作证的，能证实的 |
| **reproducible**/riprəˈdjʊsəbl/ adj. |
| 能繁殖的，可再生的，可复写的 |

## Terms

### 1. Renewable Sources

可再生能源泛指多种取之不竭的能源，严格来说，是人类有生之年都不会耗尽的能源。可再生能源不包含现时有限的能源，如化石燃料和核能。大部分的可再生能源其实都是太阳能的储存。可再生的意思并非提供十年的能源，而是百年甚至千年的。随着能源危机的出现，人们开始发现可再生能源的重要性。例如：太阳能，地热能，水能，风能，生物质能。所有人类活动的基本能源都来自太阳，透过植物的光合作用而被吸收。

**2. EVs = Electric Vehicles**

电动汽车。

**3. Heuristic Algorithm**

启发式算法。计算机科学的两大基础目标，就是发现可证明其执行效率良好且可得最佳解或次佳解的算法。而启发式算法则试图一次提供一或全部目标。例如它常能发现很不错的解，但也没办法证明它不会得到较坏的解；它通常可在合理时间解出答案，但也没办法知道它是否每次都可以这样的速度求解。有时候人们会发现在某些特殊情况下，启发式算法会得到很坏的答案或效率极差，然而造成那些特殊情况的数据组合，也许永远不会在现实世界出现。因此现实世界中启发式算法常用来解决问题。启发式算法处理许多实际问题时通常可以在合理时间内得到不错的答案。有一类的通用启发式策略称为元启发式算法，通常使用乱数搜寻技巧。他们可以应用在非常广泛的问题上，但不能保证效率。近年来随着智能计算领域的发展，出现了一类被称为超启发式算法的新算法类型。最近几年，智能计算领域的著名国际会议（GECCO 2009, CEC 2010，PPSN 2010）分别举办了专门针对超启发式算法的workshop 或 session。从 GECCO 2011 开始，超启发式算法的相关研究正式成为该会议的一个领域。国际智能计算领域的两大著名期刊 *Journal of Heuristics* 和 *Evolutionary Computation*也在 2010 年和 2012 年分别安排了专刊，着重介绍与超启发式算法有关的研究进展。

**4. Topology**

拓扑学是研究几何图形或空间在连续改变形状后还能保持不变的一些性质的学科。它只考虑物体间的位置关系而不考虑它们的形状和大小。在拓扑学里，重要的拓扑性质包括连通性与紧致性。 拓扑英文名是 Topology，直译是地志学，最早指研究地形、地貌相类似的有关学科。拓扑学是由几何学与集合论里发展出来的学科，研究空间、维度与变换等概念。这些词汇的来源可追溯至哥特佛莱德·莱布尼茨，他在 17 世纪提出"位置的几何学"和"位相分析"的说法。莱昂哈德·欧拉的柯尼斯堡七桥问题与欧拉示性数被认为是该领域最初的定理。拓扑学的一些内容早在 18 世纪就出现了，后来在拓扑学的形成中占着重要的地位。

**5. International Energy Agency**

国际能源机构亦称"国际能源署"。经济合作与发展组织的辅助机构之一。1974 年 11月成立。现有成员国 30 个，包括爱尔兰、澳大利亚、奥地利、比利时、丹麦、联邦德国、荷兰、加拿大、卢森堡、美国、挪威、葡萄牙、日本、瑞典、瑞士、土耳其、西班牙、希腊、新西兰、意大利、英国等。总部设在法国巴黎。它的宗旨是：协调各成员国的能源政策，减少对进口石油的依赖，在石油供应短缺时建立分摊石油消费制度，促进石油生产国与石油消费国之间的对话与合作。最高权力机构为理事会，由成员国部长或其他高级官员一人组成。秘书处负责处理日常事务。该机构实质上是与第三世界产油国相对抗的一个石油消费国的国际组织。成立以来，在石油市场、节能、新能源的开发利用等方面一直采取共同对策。成立日期：1974 年 2 月召开的石油消费国会议，决定成立能源协调小组来指导和协调与会国的能源工作。同年 11 月 15 日，经济合作与发展组织各国在巴黎通过了建立国际能源机构的决定。同年 11 月 18 日，16 国举行首次工作会议，签署了《国际能源机构协议》，并开始临时工作。1976 年 1 月 19 日该协议正式生效。宗旨：各成员国间在能源问题上开展合作，包括调整各成员国对石油危机的政策，发展石油供应方面的自给能力，共同采取节约石油需求的措施，加强长期合作以减少对石油进口的依赖，建立在石油供应危机时分享石油消费的制度，提供市场情报，以及促进它与石油生产国和其他石油消费国的关系等。总部在法国巴黎。

## Comprehension

### Blank Filling

1. Some of these challenges have recently been tackled by the Artificial Intelligence community, and in this paper we survey the state of the art of such AI approaches in the following EV application issues: _____, _____ and _____.

2. We can identify some key scientific dimensions of the problems that need to be tackled: _____ and _____.

3. We can identify some key engineering dimensions of the problems that need to be tackled: _____, _____ and _____.

### Content Questions

1. Some of these challenges have recently been tackled by the Artificial Intelligence community, and in this paper we survey the state of the art of such AI approaches in the following EV application issues?

2. We can identify some key scientific dimensions of the problems that need to be tackled?

3. We can identify some key engineering dimensions of the problems that need to be tackled?

## Answers

### Blank Filling

1. Energy efficient EV routing and range maximisation; Congestion management; Integrating EVs. into the Smart Grid

2. Uncertainty; Dynamism

3. Interoperability; Privacy; Real world validation

### Content Questions

1. Energy efficient EV routing and range maximisation: algo- rithms and mechanisms have been developed to route EVs. in order to minimise energy loss and maximise energy harvested during a trip; Congestion management: algorithms have been designed to manage and control the charging of the EVs., so as to minimise queues at charging points, and the discomfort to the drivers; Integrating EVs. into the Smart Grid: a number of mechanisms have been developed to schedule and control the charging of the EVs.( G2V) so that peaks and possible overloads of the electricity network may be avoided, while minimizing electricity cost.

2. Uncertainty: while several algorithms have been proposed to account for uncertainty in renewable energy production, very few tackle the uncertainty in arrival and departure times of EVs. and the load they will impose on the distribution network, as well as uncertainties in the reliability of communication systems used to coordinate collectives of EVs.; Dynamism: The state of the electricity grid, the production of renewable sources, the charging point availability, the congestion at communication and transportation networks and the number of EVs. available to provide V2G services, change quickly while a large number of EVs. are either driving or charging.

3. Interoperability: There is a need EV technologies to be able to work seamlessly and efficiently together; Privacy: In order for EVs. to be efficiently managed in terms of driving, charging and/or discharging, data on the location and the preferences of them must, in many cases, be obtained by a central mechanism; Real world validation: Currently, most of the mechanisms and the technologies related to the management of the EVs. remain at a theoretical or at a pilot deployment level, and thus, their effectiveness in a large scale deployment has not been validated.

## 参考译文 A

面对日益减少的化石燃料，以及气候变化对社会造成的日益严重的负面影响，一些国家已经启动了减少碳排放的国家计划。特别是，运输电气化被视为大幅度减少二氧化碳排放的主要途径之一。在过去几年里，电动汽车已经取得进展，到目前为止，在全世界部署了超过 18 万辆。尽管这个数只对应于 0.02%的车辆行驶在道路上，这是一个拥有超过 2000 万辆电动汽车的宏伟目标。国际能源署已经制定了到 2020 年的公路建设计划。

电动车可以大大减少二氧化碳的排放量，电动车大规模普及的关键在于使用可再生能源（如风能、太阳能）为它们充电。至关重要的是，考虑到这些能源的间歇性，需要开发作为智能电网一部分的机制，以确保这些能源在我们的能源系统中顺利整合。当电动汽车电力过剩时，可以储存能量，当电力有需求时，将这些能量反馈给电网，这可能会有所帮助。

事实上，电动汽车在运输过程中储存能源能够提高能源系统的效率。一方面，考虑到车辆每天只开一小部分时间（美国为 4%～5%），而很大一部分车辆停在停车场（美国为 90%），并且考虑到电动汽车配备大型电池，可作为停车时的存储设备（即车辆对电网（2G）方案的一部分），从而大大提高网络的存储容量。事实上，研究表明，如果美国有四分之一的汽车是电动的，这将使电网的电流存储容量增加一倍。另一方面，考虑到电动汽车数量庞大。需要每天充电（加州 40%的电动汽车车主每天行驶的路程超过了电动汽车充满电后电池的续航里程）。当需要充电时，它们可能会使网络过载。因此，在考虑电动汽车分布网络的约束条件下，需要建立能够实时管理电动汽车网对车（G2V）充电的新机制。此外，电动汽车路由系统应考虑电动汽车的能力。在刹车和/或下坡行驶时回收能量，并选择充分利用这一能力的路线。这样做，电动汽车就有可能减少充电次数，从而最大限度地扩大其范围，降低车主的成本，并将其对电网造成的峰值降至最低。

在这种背景下，研究人员开发了一些单独或集体管理电动汽车的技术和机制。例如，已经开发了一些基于 Web 的移动端应用程序，以便向电动汽车驾驶员提供关于充电站位置的信息，这些充电站所在位置有可用的充电站。此外，能源高效路由的原型系统已经开发出来，而能够在一小时内为电动汽车电池完全充电的新型充电器也变得越来越普遍。因此，虽然电动汽车的物理基础设施和技术已经取得了一些进展，但这些可能不足以管理电动汽车个人或集体行为背后的不确定性。控制电动汽车的活动，若用算法能解决问题，肯定涉及大量的异构实体（例如，电动车车主，充电点所有者，电网运营商），而每一个实体都有自己的目标、需求和动机（例如，电荷的能量、利润最大化），而他们将在高度动态的经营环境（例如变化的电动汽车数量、司机的不同意图等）中运行，同时必须处理一些不确定性因素（例如，电动汽车的未来导向等）。人工智能领域最近已经解决了其中的一些挑战，并应用到电动汽车中，这类人工智能方法的技术现状如下：

能源效率的电动汽车路线和范围最大：算法和机制已开发路由电动汽车。以尽量减少能源损失，并在旅途中获取最大的能源供给。特别是在现有搜索算法的基础上，针对电动汽车的需求和特点开发了相应的解决方案，以充分利用它们的电池恢复能力，使驾驶距离最大化。例如，提出了有或没有充电的节能电动汽车路由算法，同时提供了一种算法，在给定初始电池电量水平的某一起始点计算可到达位置。此外，利用机器学习和数据挖掘技术，加强超级电容器的使用，以尽量扩大电动汽车的范围。

拥塞管理：为管理和控制电动汽车充电而设计的算法。尽量减少在充电站排队的情况，以及令司机感到不适的情况。例如，提出了电动汽车的路由算法。减少充电点拥堵存在，考虑到司机的偏好和约束（例如，最终的目的地电力充电桩的数量），同时提出了一种启发式算法将充电点给出一定的拓扑结构，使电动车能够任意两个地点之间的旅行不会耗尽电量。

结合电动汽车，进入智能电网：开发许多新型机制为计划和控制电动汽车的充电。（G2V），以避免电网的峰值和可能的过载，同时尽量减低电力成本。此外，我们亦研究如何高效利用电动汽车的储存容量。（V2G）以平衡网络内不同地点的电力需求，或舒缓间歇性可再生能源与电网的整合。

为了在概念层次上阐明上述挑战之间的共同点和区别，我们在对研究场景进行了抽象描述。虽然我们使用树形图表示（表示概念之间的描述），但树的不同节点之间显然存在重叠（例如，在拥塞管理方面）。

因此，从这一研究现状的表现可以看出，对于电动汽车是否可以有不同的考量。首先，能否行驶取决于电池电量（即需要行驶到目的地或充电），这又给电网对车辆和车辆对电网系统在负载平衡或拥堵管理等方面带来了挑战。与此同时，激励电动汽车车主走特定路线、在特定时间充电（例如，避免峰值）或组成电动汽车集体企业在能源市场上交易的问题也随之而来。最后，还需要设计基础设施来处理大量的电动汽车。（例如，在适当地方设置充电站）以任何机制为电动汽车充电。或者把闲置的电力卖给电网。

在这种背景下，我们可以确定需要解决一些关键科学问题：

不确定性：虽然已经提出了几种算法来考虑可再生能源生产的不确定性，但很少有算法能够解决电动汽车到达和离开时间的不确定性，以及它们将给配电网带来的负荷，和用于协调电动汽车集团的通信系统可靠性的不确定性。这里的挑战必须在短时间内做出预测，因为晚做的决定可能会对运输网络造成重大破坏。因此，需要开发有效的机器学习算法来预测系统中的行为。特别是通过构建更好的人类移动模型，以及融合来自整个交通网络的数据来改善预测。为此，大规模部署电动汽车是至关重要的。未来的机器学习算法是否应该在足够大的数据集中进行训练和评估，从而使其效率最大化。

动态性：电网状态、可再生能源生产、充电站可用性、通信和交通网络拥堵、电动汽车数量、可提供的 V2G 服务、当大量电动车在行驶或充电时能够迅速应变。在这种动态设置下，需要在最小化通信带宽的同时，在短时间内解决故障安全机制和近似算法的优化问题。虽然我们已经注意到许多使用随机动态规划的解决方案，但如果能够分散这些解决方案，以确保系统更加有效，这将是有趣的。

此外，我们可以确定一些关键工程方面需要解决的问题：

互操作性：需要 EV 技术能够无缝且高效地协同工作。不同类型的充电器应该能够适用于所有电动汽车车型，实体电动汽车、充电站、网站运营商之间的数据交换应该有一个可理解的格式和含义。例如，语义 Web 技术，如 XML、RDF 和本体，可以提供一种结构化和一致的方式来表示交换的数据，从而使各种技术的协作更加有效。

隐私：为了电动汽车在驾驶、充电和/或放电方面得到有效的管理，在许多情况下，必须由一个中心机制取得关于它们的位置和喜好的数据。这就产生了隐私和数据保护的问题，因为司机可能不愿意透露这些信息。

现实世界验证：目前，大多数机制和技术都与电动汽车的管理有关，停留在理论或试验层面。因此，它们在大规模应用中的有效性尚未得到验证。设计有效的接口使人与电动车实现顺利和高效的互动，研究如何激励人按照系统给出的指令进行操作的方法，（例如，路径选择系统给出指令，提示节能的路线或充电桩的地方），这些都是至关重要的。此外，大量实体(如电动汽车、充电站、电网管理人员)协调的复杂性，以及系统对意外情况（如大量电动汽车需要在短时间内充电）的反应能力和防止负面事件（如：电网过载）必须仔细研究、分析和验证。

我们的研究表明，基于 AI 的方法正在电动汽车管理的各个领域出现：从电池充电算法到网络拥塞管理算法。我们相信，在运输工程师、电力系统专家和人工智能研究人员的共同努力下，将研发出面向实际的优秀的方案。因此，我们提倡在现场试验中与不同类型的用户联合部署更多新颖的 AI 解决方案，以解决在电动汽车大规模使用之前仍需解决的更具体的挑战。此外，人工智能技术的探索性质可以帮助电动汽车研究人员快速探索优化搜索空间由于现代人工智能基于坚实的科学方法，所有的实验和结果都是可验证的和可复制的。因此，工程师可以根据人工智能对电动汽车的研究成果，制定标准，这对于实现电动汽车活动的系统管理至关重要。目前，已有多项电动汽车相关标准出台，正在进行中的项目也多达 20 个。

# Text B

Big data is increasingly becoming a factor in production, market competitiveness and, therefore, growth. *cutting edge* analysis technologies are making inroads into all areas of life and changing our day-to-day existence. Sensor technology, *biometric* identification

**New Words and Expressions**

**cutting edge**
前沿的
**biometric**/baɪəʊˈmetrɪk/ n.
计量生物学

and the general trend towards a convergence of information and communication technologies are driving the big data movement.

Big data is increasingly becoming a factor in production, market competitiveness and, therefore, growth. Cutting edge analysis technologies are making inroads into all areas of life and changing our day-to-day existence. Sensor technology, biometric identification and the general trend towards a convergence of information and communication technologies are driving the big data movement.

Decision support for real-time traffic management is a critical component for the success of intelligent transportation systems. Theoretically, microscopic simulation models can be used to evaluate traffic management strategies in real time before a course of action is recommended. However, the problem is that the strategies would have to be evaluated in real time; this might not be computationally feasible for largescale networks and complex simulation models. To address this problem, two artificial intelligence (AI) paradigms-support vector regression (SVR) and case-based reasoning (CBR)—are presented as alternatives to the simulation models as a decision support tool. Specifically, *prototype* SVR and CBR decision support tools are developed and used to evaluate the likely impacts of implementing diversion strategies in response to incidents on a highway network in Anderson, South Carolina. The *performances* of the two prototypes are then evaluated by a comparison of their *predictions* of traffic conditions with those obtained from VISSIM, a microscopic simulation model. Although the prototype systems' predictions were comparable to those obtained by simulation, their run times were only fractions of the time required by the simulation model. Moreover, SVR performance is superior to that of CBR for most cases considered. The study results provide motivation for consideration of the proposed AI paradigms as potential decision support tools for real- time transportation management applications.

Since the early 1990s, the transportation community has turned to the use of intelligent transportation systems (ITS) to help address some of the nation's toughest transportation problems. ITS has now been deployed in almost all major U. S. cities. One of the key ITS components is the real- time traffic and incident management system. This system is designed to *optimize* the use of the existing transportation capacity, especially during incidents. Highway incidents (e.g., traffic

*New Words and Expressions*

**prototype**/ˈprəʊtətaɪp/ n.

原型，雏形，蓝本

**performance**/pəˈfɔːm(ə)ns/ n.

性能；绩效

**prediction**/prɪˈdɪkʃ(ə)n/ n.

预言，预言的事物

**optimize**/ˈɒptɪmaɪz/ vt.

使最优化，使完善

crashes, adverse weather conditions, *hazardous* material spills, and shortterm construction work) can cause excessive traffic delays and may result in secondary incidents.

The goal of incident management systems is to alleviate this kind of shortterm congestion and to smooth highway traffic flow by managing the traffic in real time. To achieve this goal, incident management systems quickly detect and verify incidents, deploy the right equipment to the scene, and attempt to manage traffic better during the incident by diverting traffic onto alternative routes, if such a diversion would help save travel time.

Whether diversion strategies may be beneficial depends on the duration and severity of the incident as well as the attractiveness of alternative routes. For example, if an incident lasts only 10 min and blocks only one lane of a three-lane highway, then diverting traffic to an alternative route that is 10. 0 mi longer is not necessarily an effective *strategy*. A key decision in the incident management process is whether diverting traffic is *warranted* for a given *scenario*.

Several *techniques* have been proposed over the years to address this question. For example, simulation models such as VISSIM, DYNA-SMART, and DynaMIT can *evaluate* traffic conditions under the diversion strategy and for a control scenario that involves no diversion. A comparison of traffic conditions under these two cases would help determine whether diversion is warranted. The problems with this approach are that running the model twice requires time-especially for largescale, complex networks and diversion strategies must be developed almost in a real-time fashion to be effective. One cannot afford to wait until the simulation runs are completed to decide on a course of action.

To address this issue, the *feasibility* of using two artificial intelligence (AI) paradigms to generalize the results obtained from running a *comprehensive* microscopic traffic simulation models is examined. If successful, these tools could then be used to evaluate new traffic situations and make routing decisions similar to those that would have been reached using a simulation model, in a fraction of the time that running a simulation model would require. It therefore would make those paradigms appropriate for online, real- time applications, and allow them to be used for real- time decision support during an incident.

---

### New Words and Expressions

**hazardous**/ˈhæzədəs/ adj.
有危险的；冒险的；碰运气的

**strategy**/ˈstrætədʒɪ/ n.
策略

**warranted**/ˈwɔrənt/ adj.
保证的；批准的

**scenario**/sɪˈnɑrɪəʊ/ n.
方案；情节

**technique**/tekˈniːk/ n.
技巧，技术；手法

**evaluate**/ɪˈvæljʊeɪt/ vt.
评价；估价

**feasibility**/fiːzɪˈbɪlɪtɪ/ n.
可行性；可能性

**comprehensive**
/kɒmprɪˈhensɪv/ adj.
综合的；广泛的

In contrast, CBR was used in earlier studies as a decision support tool for numerous ITS applications, including predicting the benefits of traffic routing on a Connecticut test network. In the current study, the feasibility of SVR and CBR in traffic management is evaluated by a comparison of their predictions of traffic state evolution against a comprehensive traffic simulation model.

## 参考译文 B

大数据正日益成为生产、市场竞争力和增长的一个因素。尖端的大数据分析技术正在渗透到生活的各个领域，并改变着我们的生活。传感器技术、生物特征识别和信息通信技术的融合正在推动大数据不断前进。

实时交通管理决策支持是智能交通系统成功的关键因素。从理论上讲，微观仿真模型可以用于交通管理策略的实时评价。在推荐一个行动方案之前，要实时地制定管理策略。然而，问题是这些策略必须实时评估；对于大规模网络和复杂的仿真模型，这在计算上可能是不可行的。为了解决这个问题，人工智能（AI）的两种范式-支持向量回归（SVR）和基于案例的推理（CBR），分别作为决策支持工具与仿真模型的替代方案。具体来说，采用原型 SVR 和 CBR 决策支持工具，来评估实施改道战略对安德森山高速公路网络发生的事故可能产生的影响，然后通过与微观模拟模型（VISSIM）预测交通状况的比较来评估这两种原型的性能。虽然原型系统的预测结果与仿真结果相当，但它的运行时间仅为仿真模型所需时间的一小部分。此外，在大多数情况下，SVR 性能优于 CBR。所以考虑将人工智能范式作为实时运输管理的决策支持工具。

自 20 世纪 90 年代以来，交通运输界开始使用智能交通系统（ITS）来帮助解决每个国家最棘手的交通问题，在美国的主要城市使用。该系统旨在优化现有运输能力，特别是在事故发生期间，公路交通事故造成的交通延误可能导致二次交通事故。

事件管理系统的目标是通过对交通的实时管理来缓解这种短期拥堵，使公路交通畅通。为了实现这一目标，事件管理自动测试系统将快速检测并核实事故，分析并向现场部署正确的设备，在事故期间更好地管理交通，如果交通改道有助于节省交通时间，则在事件发生期间，尽量将交通改道，以更有效地管理交通。

交通改道策略是否有益，取决于事件的持续时间和严重程度以及转向其他路线的效率。例如，如果事故持续仅 10 分钟，并封锁了一条三车道公路中的一条车道，然后将交通转向一条长为 10 英里的替代路线并不一定是一种有效的决策。事件管理过程中的一个关键决定是在给定的情况下是否需要交通改道。

多年来提出了若干解决这一问题的技术。例如，VISSIM、DYNASMART 和 DynaMIT 等仿真模型可以评估改道策略下的交通状况，以及不涉及改道的控制方案。比较这两种情况下的交通情况将有助于确定是否需要改道。这种方法的问题是两次运行该模型需要时间，特别是对于大规模复杂的网络，并且必须几乎以实时的方式制定转移策略。而一个人不能等到模拟运行完成后才决定行动方向。

　　为了解决这个问题，使用两种人工智能（AI）范式来概括运行一个全面的微观交通仿真模型所得到的结果是可行的。如果成功的话，这些工具就可以用于评估新的通信量和运行仿真模型所需的时间。并做出类似于使用仿真模型所能达到的路径决策。因此，它将使这些范例适合于在线实时应用程序，并允许它们在事故期间用于实时决策支持。

　　相比之下，CBR 在早期的研究中被用作众多 ITS 应用的决策支持工具，包括预测康涅狄格测试网络上流量路由的好处。本研究通过将 SVR 和 CBR 的交通状态演化预测与综合交通仿真模型进行比较，评价了 SVR 和 CBR 在交通管理中的可行性。

# Chapter *12*

# Artificial Intelligence and Home Automation

## Text A

Home automation system achieved great popularity in the last decades and it increases the comfort and quality of life. In this paper an *overview* of current and emerging home automation systems is discussed. Nowadays most home automation systems consist of a *smart phone* and *microcontroller*. A smart phone application is used to control and *monitor* the home appliances using different type of communication techniques.

Home automation system is growing rapidly, they are used to provide comfort, convenience, quality of life and security for residents. Nowadays, most home automation systems are used to provide ease to elderly and disabled people and they reduce the human labor in the production of services and goods. Home automation system can be designed and developed by using a single controller which has the ability to control and monitor different interconnected appliances such as power *plugs*, lights, temperature and *humidity* sensors, smoke, gas and fire *detectors* as well as emergency and security systems. One of the greatest advantage of home automation system is that it can be controlled and managed easily from an array of devices such as smart phone, *tablet*, *desktop* and *laptop*. The rapid growth of wireless technologies influences us to use smart phones to *remotely* control and monitor the home appliances around the world. Several home automation systems use smart phones to communicate with microcontrollers using various wireless communication techniques

### New Words and Expressions

**overview**/ˈəʊvəvjuː/ n.
综述；概观

**smart phone**
智能电话

**microcontroller**
/ˌmaikrəukənˈtrəulə/ n.
微控制器

**monitor**/ˈmɒnɪtə/ n.
监控器；显示屏

**plug**/plʌg/ n. vt.
用插头将电源接通；接插头

**humidity**/hjuˈmɪdəti/ n.
湿度；湿气

**detector**/dɪˈtektə/ n.
探测器；检测器

**tablet**/ˈtæblɪt/ n.
平板电脑

**desktop**/ˈdesktɒp/ n.
台式机

**laptop**/ˈlæptɒp/ n.
膝上型轻便电脑

**remotely**/rɪˈməʊtlɪ/ adv.
遥远地；偏僻地

such as Bluetooth, GSM, ZigBee, Wi-Fi and EnOcean. smart phone applications are used to connect to the network so that the authorized users can adjust the setting of system on their personal devices. Different type of home automation systems offer a wide range of functions and services, some of the common *features* are appliance control, thermostat control, remote control lighting, live video *surveillance*, monitor security camera, real time text alerts.

This paper describes the implementation and working principles of some existing home automation techniques and it compares their cost, speed, real time existence and other functionalities. There are different home automation technologies accessible in market but guidelines about these technology is very low, in this research work a comparison of some existing home automation technologies is discussed so users can choose their own choice of technology. This paper also discusses the comparison of some popular home automation techniques and highlight their advantages and drawbacks.

### A. Bluetooth Based Home Automation System

Home automation systems using smart phone, Arduino board and Bluetooth technology are secured and low cost. A Bluetooth based home automation system proposed by R.Piyare and M.Tazil. The hardware *architecture* of this system consists of the Arduino BT board and a cell phone, the communication between Arduino BT board and cell phone is *wirelessly* using Bluetooth technology. The Arduino BT board has a range of 10 to 100 meters, 3 Mbps data rate and 2.4 GHz *bandwidth*. In this system home appliances are connected to the Arduino BT board via relay. The cell phone use a software application which allows the user to control the home appliances. Moreover, this system used password protection to make system secure and allow only authorized user. It has the advantage to easily fit into an existing homes and automated system. The main drawback of system is that it is limited to control the home appliances within the Bluetooth range. Fig. 12-1 illustrates the block diagram of Bluetooth based home automation system. A low cost and user friendly, smart living system is presented which also use Android application to control home appliances. The wireless connection between Android device and home appliances is developed via Bluetooth technology. It also provided security and alert system for proposed smart living system.

---

### New Words and Expressions

**feature** /ˈfiːtʃə/ n.
特色；特征；容貌

**surveillance**
/səˈveɪl(ə)ns; -ˈveɪəns/ n.
监督；监视

**architecture** /ˈɑːkɪtektʃə/ n.
建筑学；建筑风格

**wirelessly**
无线地

**bandwidth** /ˈbændwɪdθ/ n.
带宽

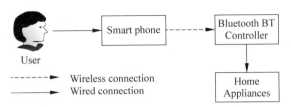

Figure 12-1  Block diagram of home automation system

### B. Voice Recognition Based Home Automation

A *voice recognition* based home automation system proposed and *implemented* by a *researcher*. The hardware architecture of this system consists of Arduino UNO and smart phone. The wireless communication between the smart phone and the Arduino UNO is done through Bluetooth technology. Android OS has a *built in* voice recognizing feature which is used to develop a smart phone application which has ability to control the home appliances from user voice command. This application converts the user voice command into text, then it transmit that text message to Bluetooth *module* HC-05 which is connected with Arduino UNO. One advantage of voice controlled home automation system is that user only pronounce the appliance name in smart phone microphone and telling it to switch ON or OFF the appliances, in this way the users can control home appliance easily without any effort. A voice recognition application provided a user friendly *interface* to users and it has ability to add more home appliances into the system. This home automation system can be used in every building using electrical appliances and devices. The main *drawback* of system is that it has limited range due to Bluetooth, its range can be extended using internet instead of Bluetooth but this solution will not be *cost effective*. This system also failed to work efficiently in a noisy environment. The block diagram of voice recognition based home automation system (HAS) is shown in Fig. 12-2, and another voice recognition based home automation system is designed by using GPRS technology. This system allows the user to control home appliances using voice commands. In this system machine learning *classifier* Support Vector Machine (SVM) is used for speech recognition.

### C. ZigBee Based Wireless Home Automation System

ZigBee based wireless home automation system has also been studied, it consists of three main modules, *handheld microphone*

*New Words and Expressions*

**voice recognition**
语音识别

**implement** /ˈɪmplɪm(ə)nt/ n. vt.
实施，执行

**researcher** /rɪˈsɜːtʃə/ n.
研究员

**built in**
安装在里面的；装入的

**module** /ˈmɒdjuːl/ n.
模块；组件

**interface** /ˈɪntəfeɪs/ n.
界面

**drawback** /ˈdrɔːbæk/ n.
缺点，不利条件

**cost effective**
有成本效益的

**classifier** /ˈklæsɪfaɪə/ n.
分类器

**handheld microphone**
手持麦克风

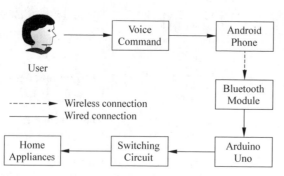

User

- - - - - - → Wireless connection
————→ Wired connection

Figure 12-2　Block diagram of the Voice controled HAS

***New Words and Expressions***

**established**/ɪˈstæblɪʃt/ adj.
　确定的；已制定的
**sample**/ˈsɑːmp(ə)l/ n.
　样品；样本
**accuracy**/ˈækjʊrəsɪ/ n.
　精确度，准确性
**accurate**/ˈækjərət/ adj.
　精确的
**consumption**/kənˈsʌm(p)ʃ(ə)n/ n.
　消费；消耗
**platform**/ˈplætfɔːm/ n.
　平台；月台

module, central controller module and appliance controller module. Handheld microphone module use ZigBee protocol and central controller module are based on PC. In this system, Microsoft speech API is used as a voice recognition application, wireless network is *established* using RF ZigBee modules due to their low power and cost efficiency. The system recorded voice at a sampling frequency of 8 kHz where as human voice highest frequency is 20 kHz. Most important part of this system is encoding which was done at frequency range between 6 Hz to 3.5 kHz. Differential pulse code modulation (DPCM) is used for compressed data from 12 bits to 6 bits. These data bits were sent from the microcontroller to the RF ZigBee module at the maximum baud rate of 115200 bits/s. The ZigBee communication protocol offers maximum baud rate of 250 kbps, but 115.2 kbps was used for microcontroller for sending and receiving data. This Automation system was tested using voice commands of 35 male and female with different English accents. Each person recorded 35 voice *samples* so total 1225 voice commands were tested and system correctly recognized 79.8% of them. Speaker accent, speed and surrounding noise affected the *accuracy* of the system. Accuracy of this system is limited in the range of 40 meters while recognition system is *accurate*, up to 80m when given a clear line of sight transmission. Fig. 12-3 illustrates the block diagram of ZigBee based home automation system.

Another ZigBee based home automation system proposed and implemented by researchers. This home automation system has two modes of operation in terms of total power *consumption*, one of them is measurement mode and second one is the current sensor mode. The Java *platform* is used for monitoring real time power.

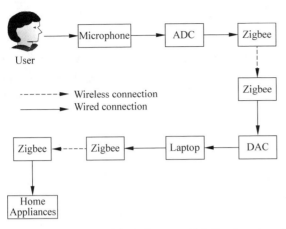

Figure 12-3   Functional block diagram of ZigBee based model

*New Words and Expressions*

**indicator** /ˈɪndɪkeɪtə/  n.

指示器

**microcontroller**

/ˌmaɪkrəʊkənˈtrəʊlə/ n.

微控制器

**feedback**/ˈfiːdbæk/ n.

反馈；成果

**coverage**/ˈkʌv(ə)rɪdʒ/ n.

覆盖范围

**provide**/prəˈvaɪd/ vt. vi.

提供；规定；准备；装备

Further, the performance of overall system has been analyzed by using different performance metrics such as Round Trip Delay (RTD) time, the Latency and Received Signal Strength *Indicator* (RSSI).

### D. GSM Based Home Automation System

A smart home automation system implemented by using Global System for Mobile communication (GSM). The hardware architecture of the system consists of GSM modem, PIC16F887 *microcontroller* and smart phone. The system used a GSM modem to control electric appliances through SMS request. PIC16F887 microcontroller interfaced with a GSM modem and it is used to read and decode the received SMS to execute the specific command. Home appliances are connected with PIC16F887 microcontroller via relays. RS232 is used for serial communication between GSM modem and PIC16F887 microcontroller. The GSM modem response time is less than 500 microseconds. The whole process of sending and receiving commands is processed within 2 seconds. One of the advantages of this automated system is that users will get *feedback* status of household appliances via SMS on their smart phones. This system was implemented in hardware and achieved 98% accuracy. Due to the wide *coverage* of GSM network users can get access to appliances from anywhere in the world. It is concluded that the usage of GSM in the home automation system *provides* maximum security and reliability. Fig. 12-4 shows the functional block diagram of GSM based home automation system (HAS).

Similarly, another GSM based home automation system designed

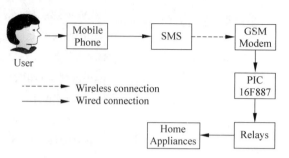

Figure 12-4    Block diagram of GSM based HAS

*New Words and Expressions*
**embed**/ɪmˈbed; em-/ vt.
　栽种；使嵌入
**gateway**/ˈɡeɪtweɪ/ n.
　门；网关；方法；通道；途径

by using the GSM SIM900 module, microcontroller LPC2148, LCD and a smart phone application for the user interface. This system enables the users to control home appliances by sending a message from Android application to GSM SIM900 module. Moreover, this system displays the important notification on the LCD and it can controlled anywhere in the world where mobile network is available.

### E. Internet of Things (IoT) Based Home Automation System

Rajeev Piyare presented a home control and monitoring system based on the internet of things (IoT) technology. It's designed and implemented by using *embedded* micro web server, controlling devices, smartphone and a software application. The architecture of system consists of three parts: home environment, home *gateway* and remote environment. Fig. 12-5 illustrates the architecture of this system.

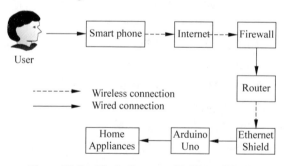

Figure 12-5    Block diagram of IoT based HAS

Remote environment allows the authorized users to remotely control and monitor the home appliances using a smartphone, which supports Wi-Fi, 3G or 4G and Android application. Home environment contains the hardware interface module and home gateway. The function of home gateway is to provide the data translation service between internet, router and Arduino Ethernet server. The most important part of home gateway is a micro Web server which is built by using an Arduino Ethernet shield. Hardware interface modules are interfaced with *actuators* and sensors via wires. This system has ability to control the energy management systems such as power plugs, lightings, security system such as gate and door locks and heating, ventilation and air *conditioning* (HVAC). For the monitoring system home environment supports sensors such as current, humanity and *temperature* sensors.

Similarly, another system focuses on controlling the home appliances through World Wide Web. This system allows the users to control and monitor the different home appliances using Wi-Fi and raspberry (server system). Home appliances such as fan, TV and light can be *remotely* controlled using the website. In addition, this system also provides protection to fire accidents and inform the user about *fireplace via* an alerting message.

### F. EnOcean Based Home Automation System

The EnOcean is newly developing energy harvesting technology used in transportation, building and home automation systems. EnOcean's technology is productively in logistics as well as in the industry due to energy efficiency and easily installing device anywhere for users ease which significantly save the installation cost up to 40%. Moreover EnOcean's devices utilize 315 MHz band and it provides convenient ways for home automation system. EnOcean based home automation system can be built up using Internet, router, automation controller, duckbill 2 EnOcean and EnOcean devices. Duckbill 2 EnOcean is thumb drive used for home automation system and it has an EnOcean TCM310 transceiver and Ethernet. Moreover Duckbill 2 EnOcean run applications under Linux system. It can work as a *stand-alone* automation controller inside building automation systems. An EnOcean based home automation system is depicted in Fig. 12-6.

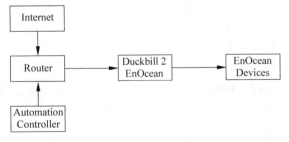

Figure 12-6　Block diagram of EnOcean based automation system

**New Words and Expressions**

**actuator** /ˈæktjʊeɪtə/ n.
　执行机构；激励者；促动器
**conditioning**/kənˈdɪʃənɪŋ/ n.
　调节；条件作用
**ventilation** /ˌventɪˈleɪʃ(ə)n/ n.
　通风设备；空气流通
**temperature sensor**
　温度传感器
**remotely**/rɪˈməʊtlɪ/ adv.
　遥远地；偏僻地
**fireplace**/ˈfaɪəpleɪs/ n.
　壁炉
**via**/ˈvaɪə,ˈviːə/
　渠道，通过；经由
**stand-alone**/ˈstændəˌləʊn/ adj.
　独立的；独立操作的

The future of home automation system requires to make homes smarter and more convenient. For future work it is suggested to develop image processing based home automation system using the above discussed technologies. In such automation system home appliances will be controlled by different gestures which will be detected through the camera. Moreover, home automation system can be developed by interfacing biomedical signals such as *Electromyography* (EMG) signal with computer, it will provide opportunity to *amputee* to control appliances from different arm gestures. It can be useful in robotics area for controlling robot through gesture for different tasks.

**New Words and Expressions**
**electromyography**
/ɪˌlɛktrəʊmaɪˈɒɡrəfi/ n.
肌电图
**amputee**/æmpjʊˈtiː/ n.
被截肢者

## Terms

### 1. IBM Watson

IBM Watson 认知计算系统的杰出代表，也是一个技术平台。认知计算代表一种全新的计算模式，它包含信息分析，自然语言处理和机器学习领域的大量技术创新，能够助力决策者从大量非结构化数据中揭示非凡的洞察。

### 2. Bluetooth

蓝牙（Bluetooth）技术是一种无线技术标准，可实现固定设备、移动设备和楼宇个人域网之间的短距离数据交换（使用 2.4~2.485GHz 的 ISM 波段的 UHF 无线电波）。蓝牙技术最初由电信巨头爱立信公司于 1994 年创制，当时是作为 RS-232 数据线的替代方案。蓝牙可连接多个设备，克服了数据同步的难题。如今蓝牙由蓝牙技术联盟管理。蓝牙技术联盟在全球拥有超过 25,000 家成员公司，它们分布在电信、计算机、网络和消费电子等多重领域。IEEE 将蓝牙技术列为 IEEE 802.15.1，但如今已不再维持该标准。蓝牙技术联盟负责监督蓝牙规范的开发，管理认证项目，并维护商标权益。制造商的设备必须符合蓝牙技术联盟的标准才能以"蓝牙设备"的名义进入市场。蓝牙技术拥有一套专利网络，可发放给符合标准的设备。

### 3. GSM

Global System For Mobile Communications（GSM），由欧洲电信标准组织 ETSI 制定的一个数字移动通信标准，GSM 是全球移动通信系统的简称。它的空中接口采用时分多址技术。自 20 世纪 90 年代中期投入商用以来，被全球超过 100 个国家采用。GSM 标准的设备占据当前全球蜂窝移动通信设备市场 80%以上。全球移动通信系统 Global System for Mobile Communication 就是众所周知的 GSM，是当前应用最为广泛的移动电话标准。全球超过 200 个国家和地区超过 10 亿人正在使用 GSM 电话。GSM 标准的无处不在使得在移动电话运营商之间签署"漫游协定"后用户的国际漫游变得很平常。GSM 较之它以前的标准最大的不同是它的信令和语音信道都是数字式的，因此 GSM 被看作是第二代（2G）移动电话系统。这说明数字通信从很早就已经构建到系统中。GSM 是一个当前由 3GPP 开发的开放标准。2015 年，全球诸多 GSM 网络运营商，已经将 2017 年确定为关闭 GSM 网络的年份。

#### 4. ZigBee

ZigBee 是基于 IEEE 802.15.4 标准的低功耗局域网协议。根据国际标准规定，ZigBee 技术是一种短距离、低功耗的无线通信技术。这一名称（又称紫蜂协议）来源于蜜蜂的八字舞，由于蜜蜂（bee）是靠飞翔和"嗡嗡"（zig）地抖动翅膀的"舞蹈"来与同伴传递花粉所在方位信息，也就是说蜜蜂依靠这样的方式构成了群体中的通信网络。其特点是近距离、低复杂度、自组织、低功耗、低数据速率。主要适合用于自动控制和远程控制领域，可以嵌入各种设备。简而言之，ZigBee 就是一种便宜的，低功耗的近距离无线组网通信技术。ZigBee 是一种低速短距离传输的无线网络协议。ZigBee 协议从下到上分别为物理层（PHY）、媒体访问控制层（MAC）、传输层（TL）、网络层（NWK）、应用层（APL）等。其中物理层和媒体访问控制层遵循 IEEE 802.15.4 标准的规定。

#### 5. Wi-Fi

无线网络（Wi-Fi）在无线局域网的范畴是指"无线相容性认证"，实质上是一种商业认证，同时也是一种无线联网技术，以前通过网线连接计算机，而 Wi-Fi 则是通过无线电波来连网；常见的就是一个无线路由器，那么在这个无线路由器的电波覆盖的有效范围都可以采用 Wi-Fi 连接方式进行联网，如果无线路由器连接了一条 ADSL 线路或者别的上网线路，则又被称为热点。

#### 6. EnOcean

EnOcean 是一种基于能量收集的超低功耗短距离无线通信技术，被应用于室内能量收集，在智能家居、工业、交通、物流也有应用。基于 EnOcean 技术的模块有高质量无线通信、能量收集和转化及超低功耗的特点。其通信协议非常精简，采用无须握手的通信机制，相较于其他无线通信技术如 ZigBee 有更低的功耗和更高的效率；EnOcean 还可以通过收集自然界的微小能量为模块提供能源，使模块做到无电池和免维护。在 2012 年 3 月，EnOcean 无线技术被国际标准 ISO/IEC 14543-3-10 批准认证。标准覆盖 EnOcean 网络体系结构的前三层：物理层、数据链路层和网络层。第四层由 EnOcean 开放联盟来负责制定。

## Comprehension

### Blank Filling

1. Home automation systems using smartphone, Arduino board and Bluetooth technology are _____ and _____.
2. ZigBee based wireless home automation system has also been studied, it consists of three main modules, _____, _____ and _____.

### Content Questions

1. What is Bluetooth based home automation system?
2. The main drawback of Voice recognition based home automation?

## Answers

### Blank Filling

1. secured; low cost

2. handheld microphone module; central controller module; appliance controller module

**Content Questions**

1. In this system home appliances are connected to the Arduino BT board via relay. The cell phone uses a software application which allows the user to control the home appliances. Moreover, this system used password protection to make system secure and allow only authorized user. It has the advantage to easily fit into an existing homes and automated system. The main drawback of system is that it is limited to control the home appliances within the Bluetooth range. Fig. 12.1 illustrates the block diagram of Bluetooth based home automation system. A low cost and user friendly, smart living system is presented which also use Android application to control home appliances. The wireless connection between Android device and home appliances is developed via Bluetooth technology. It also provided security and alert system for proposed smart living system.

2. The main drawback of system is that it has limited range due to Bluetooth, its range can be extended using internet instead of Bluetooth but this solution will not be cost effective. This system also failed to work efficiently in a noisy environment.

## 参考译文 A

　　家庭自动化系统发展迅速，为居民提供舒适、方便、优质的生活和安全保障。目前，大多数家庭自动化系统都是为老年人和残疾人提供方便，减少了服务和商品生产中的人工劳动。家庭自动化系统可以通过一个控制器来设计和开发，该控制器能够控制和监控不同的相互连接的设备，如电源插头、灯、温湿度传感器、烟雾、气体和火灾探测器以及应急和安全系统。家庭自动化系统最大的优点之一是，它可以通过智能手机、平板电脑、台式机和笔记本电脑等一系列设备轻松控制和管理。无线技术的快速发展使我们可以使用智能手机远程控制和监控世界各地的家用电器。一些家庭自动化系统使用智能手机与微控制器通信，使用各种无线通信技术，如蓝牙、GSM、ZigBee、Wi-Fi 和 EnOcean。智能手机应用程序连接到网络上，授权用户可以在自己的个人设备上调整系统设置。不同类型的家庭自动化系统提供了广泛的功能和服务，一些共同的特点是电器控制，恒温控制，远程控制照明，实时视频监控，监控安全摄像头，实时文本提醒。

　　我们介绍了一些现有的家庭自动化技术的实现和工作原理，并对其成本、速度、实时存在性等功能进行了比较。市场上有各种各样的家庭自动化技术，但是关于这些技术的指导方针非常低，在本研究工作中，我们对一些现有的家庭自动化技术进行了比较，以便用户可以自己进行选择。本文还对几种常用的家庭自动化技术进行了比较，指出了各自的优缺点。

### 1. 基于蓝牙的家庭自动化系统

　　家庭自动化系统使用智能手机，Arduino 板和蓝牙技术是安全和低成本的。提出了一种基于蓝牙的家庭自动化系统 Piyare M.Tazil。本系统硬件架构由 Arduino BT 板和手机组成，Arduino BT 板与手机之间的通信采用蓝牙无线技术。Arduino BT 板具有 10～100m 的范围，3Mb/s 的数据速率和 2.4GHz 的带宽。在该系统中，家用电器通过继电器连接到

Arduino BT 板上。手机使用一个软件应用程序，它允许用户控制家电。此外，本系统采用密码保护，使系统安全，只允许授权用户使用。它的优点是可以很容易地适应现有的家庭和自动化系统。该系统的主要缺点是将家电控制在蓝牙范围内。图 12-1 为基于蓝牙的家庭自动化系统框图。提出了一种低成本、用户友好、智能的生活系统，并利用 Android 应用程序对家电进行了控制。Android 设备与家电之间的无线连接是通过蓝牙技术开发的。并为提出的智能生活系统提供了安全预警系统。

### 2. 基于语音识别的家庭自动化

AI 工程师提出并实现了一种基于语音识别的家庭自动化系统。该系统的硬件架构由 Arduino UNO 和智能手机组成。智能手机和 Arduino UNO 之间的无线通信是通过蓝牙技术完成的。Android 操作系统有一个内建的语音识别功能，用来开发一个智能手机应用程序，可以通过用户语音命令来控制家电。该应用程序将用户语音命令转换为文本，然后将该文本消息传输到与 Arduino UNO 连接的蓝牙模块 HC- 05。语音控制家居自动化系统的一个优点是用户只需在智能手机麦克风中读出家电名称，然后告诉它开启或关闭家电，这样用户就可以毫不费力地轻松控制家电。语音识别应用程序为用户提供了友好的用户界面，能够在系统中添加更多的家用电器。家庭自动化系统可以使用电器和设备在每一个建筑物中使用。系统的主要缺点是由于蓝牙的限制，范围有限，可以用互联网代替蓝牙扩展范围，但是这种解决方案并不划算。该系统在噪声环境下也不能有效工作。基于语音识别的家庭自动化系统框图如图 2 所示，利用 GPRS 技术设计了另一个基于语音识别的家庭自动化系统。该系统允许用户使用语音命令控制家用电器。该系统采用机器学习分类器支持向量机（SVM）进行语音识别。

### 3. 基于 ZigBee 的无线家庭自动化系统

基于 ZigBee 的无线家庭自动化系统，主要由手持麦克风模块、中央控制器模块和家电控制器模块三个模块组成。手持话筒模块采用 ZigBee 协议，中央控制器模块基于 PC。本系统采用 Microsoft 语音 API 作为语音识别应用，采用射频 ZigBee 模块建立无线网络，具有功耗低、成本低的特点。系统以 8kHz 采样频率记录语音，其中人类语音最高频率为 20kHz。该系统最重要的部分是编码，编码的频率范围在 6Hz～3.5kHz 之间。差分脉冲码调制（DPCM）用于从 12 位到 6 位的压缩数据。这些数据位从单片机以 115200b/s 的最大波特率发送到 RF ZigBee 模块。ZigBee 通信协议最大波特率为 250kb/s，单片机采用 115.2kb/s 进行数据收发。该自动化系统使用 35 个不同英语口音的男性和女性语音命令进行测试。每个人记录 35 个语音样本，共测试 1225 个语音命令，系统正确识别 79.8%。扬声器的口音，速度和周围的噪音影响了系统的准确性。该系统的精度限制在 40m 范围内，而识别系统是准确的，当给定一个清晰的视线传输时，最高可达 80m。图 12-3 为 ZigBee 家庭自动化系统框图。

研究人员提出并实现了另一种基于 ZigBee 的家庭自动化系统。该家庭自动化系统按总功耗分为两种运行模式，一种是测量模式，另一种是电流传感器模式。Java 平台用于实时监控电源。此外，通过使用往返延迟（RTD）时间、延迟和接收信号强度指示器（RSSI）等不同的性能指标分析了整个系统的性能。

### 4. 基于 GSM 的家庭自动化系统

采用全球移动通信系统（GSM）实现的智能家居自动化系统。系统硬件结构由 GSM

调制解调器、PIC16F887 单片机和智能手机组成。该系统采用 GSM 调制解调器通过短信请求对电器进行控制。PIC16F887 单片机与 GSM 调制解调器接口，用于读取和解码接收到的短信，执行特定的命令。家用电器通过继电器与 PIC16F887 单片机连接。RS-232 用于 GSM 调制解调器与 PIC16F887 单片机之间的串行通信。GSM 调制解调器的响应时间小于 500μs。整个发送和接收命令的过程在 2s 内完成。该自动化系统的优点之一是用户可以通过智能手机上的短信获得家电的反馈状态。这个系统是在硬件中实现的，并取得了 98%的准确率。由于 GSM 网络的广泛覆盖，用户可以从世界任何地方访问设备。结果表明，GSM 在家庭自动化系统中的应用提供了最大的安全性和可靠性。图 12-4 为基于 GSM 的家庭自动化系统功能框图（HAS）。

类似的，另一个基于 GSM 的家庭自动化系统设计采用了 GSM SIM900 模块，单片机 LPC2148, LCD 和一个智能手机应用程序为用户界面。本系统通过向 GSM SIM900 模块发送一条来自 Android 应用程序的消息，实现对家电的控制。此外，该系统还能将重要的通知显示在液晶显示屏上，并可在世界上任何一个移动网络可用的地方进行控制。

**5. 基于物联网的家居自动化系统**

RajeevPiyare 介绍了一种基于物联网技术的家庭监控系统。它是通过嵌入式微 Web 服务器、控制设备、智能手机和一个软件应用程序来设计和实现的。系统架构由家庭环境、家庭网关和远程环境三部分组成。图 12-5 为本系统架构示意图。

远程环境（控制系统）允许授权用户使用支持 Wi-Fi、3G 或 4G 以及 android 应用的智能手机远程控制和监控家电。家庭环境包含硬件接口模块和家庭网关。家庭网关的功能是提供互联网、路由器和 Arduino 以太网服务器之间的数据翻译服务。家庭网关最重要的部分是使用 Arduino 以太网屏蔽构建的微型 Web 服务器。硬件接口模块通过导线与执行机构和传感器接口。本系统能够控制电源插头、照明、门禁、门锁等能源管理系统和采暖、通风、空调（HVAC）等安全系统。对于监控系统，家庭环境支持电流、湿度和温度等传感器。

同样，另一个系统专注于通过万维网来控制家用电器。本系统允许用户使用 Wi-Fi 和 raspberry（服务器系统）对不同的家电进行控制和监控。家用电器如风扇、电视和电灯可以通过这个网站进行远程控制。此外，该系统还可以对火灾事故进行保护，并通过报警信息告知用户关于壁炉的情况。

**6. 基于 EnOcean 的家庭自动化系统**

EnOcean 公司开发一种用于交通运输、建筑和家庭自动化系统的新能源收集技术。EnOcean 的技术在物流和行业中都是富有成效的，因为它的能源效率高，而且可以方便地在任何地方安装设备，方便用户，大大节省了高达 40%的安装成本。EnOcean 的设备采用 315MHz 频段，为家庭自动化系统提供了方便的方式。利用 Internet、路由器、自动化控制器、duckbill 2、EnOcean 设备可以建立基于 EnOcean 的家庭自动化系统。Duckbill 2 EnOcean 是家用自动化系统的拇指驱动器，例如有 EnOcean TCM310 收发机和以太网。此外，Duckbill 2 在 Linux 系统下运行应用程序。它可以作为楼宇自动化系统中的独立自动化控制器。一个基于 EnOcean 的家庭自动化系统如图 12-6 所示。

未来的家庭自动化系统要求使家庭更智能、更方便。在今后的工作中，建议利用上述技术开发基于图像处理的家庭自动化系统。在这种自动化系统中，家用电器将由不同的手势控制，这些手势将通过摄像机检测出来。此外，还可以通过将肌电图（EMG）信号等生物医学信号与计算机接口，开发出家庭自动化系统，为截肢患者提供从不同的手臂姿态控制设备的机会。它可以用于机器人领域，通过手势控制机器人的不同任务。

# Text B

Smart homes have *emerged* as an area of research interest in recent years. Home automation offers us *independence* and complete control over your home along with a fine designed intelligent architecture which could control your home in your *absence in* an efficient manner. In the recent years all the implemented techniques have not realized the Intelligent Home design in all quality aspects since every technique has its own pros and cons whether we talk in terms of technology adapted, efficiency or cost. This paper presents a brief comparative analysis on implemented techniques and provides a viable solution to realize home automation system which constitutes Bluetooth control via Android app development for *in-house* control and GSM (Global System for Mobile Communication) technology for mobile control using Arduino Development Board as brain of our system. To implement Intelligent Home, the need is to adapt simple, efficient and cost effective technologies and the solution presented in this paper constitutes the said features.

Home automation has been one of the most recognizable technologies that has been *utilized* by both industrial and private *sectors* of life in the 21st century. Various definitions of Home Automation have been presented and with the ever advancing technology these definitions are improving on daily basis. explains Home Automation as a pleasing technology that may be introduced for the *inhabitants* of a house to improve the way they spent their life in-house and hence taking the living standards to another level by introducing facilitations as efficient utilization of energy resources, improved multimedia experience and delivering health services through information technology. This technology has been advancing using incorporation of new techniques to ensure reliability and addition of new features. Home automation is a technology involving *centralized* & autonomous control of housing, buildings and industry, including safety features against various sudden unanticipated scenarios. Home automation basically incorporates an electronic control of

| **New Words and Expressions** |
| :--- |
| **emerge** /ɪˈmɜːdʒ/ vi. |
| 浮现；摆脱；暴露 |
| **independence**/ɪndɪˈpend(ə)ns/ n. |
| 独立性，自立性；自主 |
| **absence in** |
| 没有在 |
| **in-house**/ɪn-haʊs/ adv. adj. |
| 内部地；内部的 |
| **utilized**/juːtəlaɪzd/ |
| 被利用的 |
| **sector**/ˈsektə/ n. |
| 部门；扇形 |
| **inhabitant**/ɪnˈhæbɪt(ə)nt/ n. |
| 居民；居住者 |
| **centralize**/ˈsentrəlaɪz/ vt. vi. |
| 使成为…的中心；集中 |

household activities like control of electrical appliances, lightning, central heating & air conditioning and security system. These features are assisted by using different types of sensors, control and communication modules. Various home automation techniques have been implemented and were presented in the form of research in recent years. All of presented techniques have their own distinct advantages over each other. In this paper we have critically analyzed techniques which are already implemented in terms of their efficiency, ease of use, *robustness* and cost. Now, some experts have propose a solution which is efficient enough to be an ideal home automation system and comply with above mentioned factors. The project integrates a centralized automation system including control & communication features along-with security modules interfaced to each other through a main master control providing human interface. The aim can be met by *integrating* GSM (AT Commands) for mobile control, ZigBee for centralized control within home and Arduino Development Board for catering GSM and ZigBee. This method results in a solution which is highly efficient, intelligent, and easy to use which could also prove a right choice for customer in terms of cost.

***New Words and Expressions***
**robustness**/rəʊˈbʌstnɪs/ n.
稳健性；健壮性
**integrating**/ˈɪntɪɡreɪtɪŋ/ n.
集成化

## 参考译文 B

近年来，智能家居已经成为一个研究热点。家庭自动化为我们提供了独立和完整的控制系统，而一个精心设计的智能架构，可以有效地控制您不在家时的家庭自动化系统。近年来，无论是从技术的适应性、效率还是成本上来说，每一种技术都有其优缺点，并没有在所有的质量方面实现智能家居设计。本文简要比较分析技术和实现，提供了一个可行的解决方案来实现家庭自动化系统，基于蓝牙技术和 Android 系统来开发出 Arduino 开发板，并通过内部控制和 GSM（全球移动通信系统）技术使我们可以通过移动终端来控制智能家居。要实现智能家居，需要适应简单、高效、低成本的技术，本文提出的解决方案构成了上述特点。

家庭自动化是 21 世纪工业和私人生活部门使用的最知名的技术之一。研究人员提出各种各样的家庭自动化的定义，随着技术的不断进步，这些定义每天都在改进。为了解释家庭自动化的技术，定义指出，家庭自动化是一项令人愉悦的技术，可以全家庭引入，以改善他们的生活方式，使其生活水平提升到一个更高的水平。能源资源可得到有效利用，提高多媒体体验，并提供卫生服务。该技术一直在推进使用新技术的结合，以确保可靠性和新功能的添加。家庭自动化是一项涉及住宅、建筑和工业的集中和自主控制的技术，包括针对各种突发意外情况的安全特性。家庭自动化主要包括家电控制、雷电控制、集中供热空调控制、安全系统控制等家庭活动的电子控制。这些功能通过使用不同类型的传感器、控制和通信模块来辅助。近年来，各种家庭自动化技术得到了应用，并以研究的形式呈现出来。所有提出的技术都有其独特的优点。在本文中，我们从效率、易用性、鲁棒性和成本等方面对已经实现的技术进行了批判性的分析。现在，一些专家已经提出了一个解决方案，这个方案是足够有效的，是一个理想的家庭自动化系统，并符合上述预期。该项目集成了一个集中的自动化系统，包括控制和通信功能，以及通过提供人机界面的主控制相互连接的安全模块。将 GSM（AT 命令）用于移动控制，ZigBee 用于家庭集中控制，Arduino 开发板用于餐饮 GSM 和 ZigBee，可以实现这一目标。该方法是一种高效、智能、易于使用的解决方案，也是客户在成本上的正确选择。

# Chapter *13*

## AI's Impact

## Text A

Artificial intelligence has impacted many aspects of human life. Today, artificial intelligence has a big impact on economic theory. In particular we study the impact of artificial intelligence on the theory of bounded rationality, efficient market hypothesis and *prospect* theory.

Artificial intelligence is a *paradigm* where computers or machines are designed to perform tasks that require high level *cognition*. This is normally achieved by looking at nature and designing machines that are inspired by objects or systems from nature that have been perfected over a long period of time. For example, one can look at how a *colony* of ants find a shortest distance from its home and the food source and use this to design routing *algorithms* that are essential for our GPS guides in our cars. The impact of artificial intelligence on major areas of economic *sectors* is extensive. In the *manufacturing* industry the application of artificial intelligence to perform tasks that used to be performed by humans will result in extensive job losses. Artificial intelligence has found applications in complex areas in the social, *political* and economic spaces.

Some scholars applied artificial intelligence extensively to model militarized interstate conflict. In this regard the problem of conflict resolution which traditionally required human *intuition* now involves using computers *empowered* with artificial intelligence to secure peace. Another application of artificial is the application of

***New Words and Expressions***

**prospect**/ˈprɒspekt/ n.

前途；预期；景色

**paradigm**/ˈpærədaɪm/ n.

范例

**cognition**/kɒgˈnɪʃ(ə)n/ n.

认识；知识

**colony**/ˈkɒlənɪ/ n.

殖民地；移民队

**algorithm**/ˈælgərɪð(ə)m/ n.

算法，运算法则

**sector**/ˈsektə/ n.

部门；扇形，扇区

**manufacturing**
/ˌmænjʊˈfæktʃərɪŋ/ adj

制造的；制造业的

**political**/pəˈlɪtɪk(ə)l/ adj.

政治的；党派的

**intuition**/ɪntjʊˈɪʃ(ə)n/ n.

直觉；直觉力

**empower**/ɪmˈpaʊə; em-/ vt.

授权，允许；使能够

artificial intelligence to better design complex *structures* such as aircrafts. In this regard some scholars were able to use artificial intelligence to create models that are *essential* to the design of complex systems such as aircraft. In decision making one essential aspect is to be able to secure all information required to make a rational decision. Artificial intelligence has been applied successfully to fill in the gaps that exist in information required to make informed decision. Some scholars applied artificial intelligence to fill in missing information and applied this to decision making in assessing the risks associated with making decisions with incomplete information. Monitoring the conditions of structures such as bridges is essential for *securing* safe *utilization* of essential public goods such as bridges. In this regard applied artificial intelligence to monitor the conditions of essential mechanical and electrical engineering structures essential in the electricity *industry*.

One *aspect* of the artificial intelligence is how this technology changes economic theories. Some scholars applied artificial intelligence to model economic and financial instruments such as the stock markets, derivatives and options. How then does artificial intelligence changes economic theory? For example, Economics Nobel Laureate Herbert Simon observed that on making decisions *rationally* one does not have the perfect and complete information to make a fully rational decision. Moreover, one does not have the perfect brain to process such information timely and efficiently and the human brain is not consistent and thus decisions made by a human brain are thus *inconsistent* as they change depending on *other factors* such as moods *swings*. Simon *termed* decision making under such *circumstances* bounded rationally. With the advent of artificial intelligence one is able to access information that was hidden and thus not accessible, and is able to use such information consistently by the use of artificial intelligence for decision making and is able to increasingly make such decisions more timely and efficiently due to Moore's Law which states that the processing power of machines is always increasing. What does this advent of artificial intelligence mean for the theory of bounded *rationality*? It means that the bounds in Simon's theory of bounded rationality are in effect *flexible* due to Moore's Law.

Another economic theory which is influenced by *the advent of* artificial intelligence is the theory of the efficient market hypothesis

**New Words and Expressions**

**structure**/ˈstrʌktʃə/ n.
结构；构造；建筑物

**essential**/ɪˈsenʃ(ə)l/ adj.
基本的；必要的

**securing**/sɪˈkjʊr/ adj.
固定住的

**utilization**/ˌjuːtɪlaɪˈzeɪʃən/ n.
利用，使用

**aspect**/ˈæspekt/ n.
方面；方向

**rationally**/ˈræʃnəli/ adv.
理性地；讲道理地

**inconsistent**/ɪnkənˈsɪst(ə)nt/ adj.
不一致的；前后矛盾的

**other factors**
其他因素

**swing**/swɪŋ/ vt. vi.
摇摆；转向

**term**/tɜːm/ n. vt.
术语；学期；期限；条款；
把…叫做

**circumstance**/ˈsɜːkəmstəns/ n.
环境，情况；事件；境遇

**rationality**/ˌræʃəˈnælɪtɪ/ n.
合理性；合理的行动

**flexible**/ˈfleksɪb(ə)l/ adj.
灵活的；柔韧的

**the advent of**
…的出现

developed by Nobel Laureate Eugene Fama. This hypothesis states that it is often difficult to beat the markets because the markets are efficient. The problem is that because the *traders* in the market are often not perfect and the information they have is imperfect and *incomplete* the markets are not efficient. Now what happens to the efficient market theory if the traders in the market are not just people but are a combination of people and artificial intelligence infused computer trader? The more artificial intelligence empowered computer traders we have in the markets the more efficient the markets become and therefore the degree at which markets are efficient depends on the amount of artificial intelligent traders we have in the markets.

Nobel Laureate Daniel Kahneman and Amos Tversky proposed the prospect theory that states that when people make decisions with the probability of *outcomes* known they weigh potential losses against potential gains to make such decisions. The impact of this theory on the markets is extensive. However, it rests on the fact that the decision maker is just a person. What happens to this theory if the decision maker is not just a person but a person who is using an artificial intelligent decision machine? How about if the decision maker is wholly an artificial intelligent machine? The applicability of Prospect Theory solely depends on how much artificial intelligent machine is used to make such a decision.

Decision making is more and more involving artificial intelligent machine. This paper described how three economic theories are impacted by the application of artificial intelligent machine in the decision making process. It is found that the use of artificial intelligent machine changes the degrees in which the theory of bounded rationality, efficient market hypothesis and prospect theories are applicable.

| *New Words and Expressions* |
| --- |
| **trader**/ˈtreɪdə/ n. |
| 　交易者；商人；商船 |
| **incomplete**/ɪnkəmˈpliːt/ adj. |
| 　不完全的 |
| **outcome**/ˈaʊtkʌm/ n. |
| 　结果，结局；成果 |
| **bounded**/ˈbaʊndɪd/ adj. |
| 　有界的；有界限的 |

## Terms

### 1. GPS

利用 GPS（Global Position System）定位卫星，在全球范围内实时进行定位、导航的系统，称为全球卫星定位系统。它是由美国国防部研制建立的一种具有全方位、全天候、全时段、高精度的卫星导航系统，能为全球用户提供低成本、高精度的三维位置、速度和精确定时等导航信息，是卫星通信技术在导航领域的应用典范，它极大地提高了地球社会的信息化水平，有力地推动了数字经济的发展。GPS 可以提供车辆定位、防盗、反劫、行驶路线监控及呼叫指挥等功能。要实现以上所有功能必须具备 GPS 终端、传输网络和监控平台三个要素。

**2. 经济领域**

经济领域指的是经济范畴，包含经济活动、经济行业以及经济理论，是经济事务的总体抽象概括。客观存在的经济关系的理论表现。它是反映某种社会生产关系的本质的概念，是从现实的经济关系中抽象出来的。人们在社会实践中得到的关于社会经济现象的感性认识，经过科学的抽象，提高到理性认识，而形成经济范畴。经济范畴同普通的经济概念不同，后者只是反映某些个别的经济现象或经济过程，经济范畴则是反映大量出现的经济现象或经济过程。根据经济范畴所反映的生产关系的内容来看，一般可分为三类：①人类一切社会共有的最一般的经济范，如生产、交换、分配、消费、生产力、生产关系等。②人类几个社会共有的经济范畴，如商品、价值、货币等。③只存在于某一个社会的特殊经济范畴，如资本、剩余价值是资本主义社会的经济范畴，按劳分配是社会主义社会的经济范畴。由于经济范畴是在一定生产关系的基础上产生的，因而"也同它们所表现的关系一样，不是永恒的。它们是历史的暂时的产物"。人类一切社会共有的最一般的经济范畴，在不同的社会形态和经济条件下，会具有不同的性质和内涵。人类几个社会共有的经济范畴和某一个社会特有的经济范畴更具有历史的暂时的性质，当生产关系发生变化时，旧的经济范畴就必然要让位于新的经济范畴，或者虽然存在旧的经济范畴的形式，但已赋予性质不同的新的内容。

**3. Theory of Bounded Rationality**

有限理性模型是指 20 世纪 50 年代之后，人们认识到建立在"经济人"假说之上的完全理性决策理论只是一种理想模式，不可能指导实际中的决策。赫伯特·西蒙提出了满意标准和有限理性标准，用"社会人"取代"经济人"，大大拓展了决策理论的研究领域，产生了新的理论——有限理性决策理论。有限理性模型又称西蒙模型或西蒙最满意模型。这是一个比较现实的模型，它认为人的理性是处于完全理性和完全非理性之间的一种有限理性。

## Comprehension

**Blank Filling**

1. Big data is being generated by everything around us at all times. Every_____ and _____ produces it.

2. _____is a paradigm where computers or machines are designed to perform tasks that require high level cognition.

3. Artificial intelligence has found applications in complex areas in the _____, _____ and _____.

4. _____ is more and more involving artificial intelligent machine.

5. It is found that the use of artificial intelligent machine changes the degrees in which the _____, _____ and _____ are applicable.

**Content Questions**

1. How then does artificial intelligence changes economic theory?

2. What does this advent of artificial intelligence mean for the theory of bounded rationality?

3. Now what happens to the efficient market theory if the traders in the market are not just people but are a combination of people and artificial intelligence infused computer trader?

## Answers

**Blank Filling**

1. Artificial intelligence

2. social; political; economic spaces

3. Decision making

4. theory of bounded rationality; efficient market hypothesis; prospect theories

**Content Questions**

1. For example, Economics Nobel Laureate Herbert Simon observed that on making decisions rationally one does not have the perfect and complete information to make a fully rational decision. Moreover, one does not have the perfect brain to process such information timely and efficiently and the human brain is not consistent and thus decisions made by a human brain are thus inconsistent as they change depending on other factors such as moods swings. Simon termed decision making under such circumstances bounded rationally. With the advent of artificial intelligence one is able to access information that was hidden and thus not accessible, and is able to use such information consistently by the use of artificial intelligence for decision making and is able to increasingly make such decisions more timely and efficiently due to Moore's Law which states that the processing power of machines is always increasing.

2. It means that the bounds in Simon's theory of bounded rationality are in effect flexible due to Moore's Law.

3. The more artificial intelligence empowered computer traders we have in the markets the more efficient the markets become and therefore the degree at which markets are efficient depends on the amount of artificial intelligent traders we have in the markets.

## 参考译文 A

人工智能已经影响了人类生活的许多方面。今天，人工智能对经济理论产生了巨大的影响。特别研究了人工智能对有限理性理论、有效市场假说和前景理论的影响。

人工智能是一种计算机或机器被设计用来执行需要高级认知能力的任务的范式。这通常是通过观察自然和设计机器来实现的，这些机器的灵感来自于经过长期完善的自然物体或系统。例如，我们可以观察一群蚂蚁如何找到家和食物来源之间最近的距离，如果利用这一点来设计路由算法，这对于我们车载 GPS 导引至关重要。人工智能对经济各主要领域的影响是广泛的。在制造业中，应用人工智能来完成过去由人类完成的任务将导致大量的失业。人工智能已经在社会、政治和经济的复杂领域中得到了应用。

一些学者将人工智能广泛应用于模拟国家间军事化冲突。在这方面，传统上需要人类直觉才能解决的冲突解决问题现在涉及使用具有人工智能的计算机来确保和平。人工智能的另一个应用是飞机等复杂结构的设计。在这方面，一些学者能够使用人工智能来设计飞机等复杂系统的模型。在决策过程中，一个重要的方面是能够确保做出合理决策所需的所有信息的安全。人工智能已被成功地应用于填补在做出明智决策所需的信息方面的空白。有学者将人工智能应用于信息缺失的填补，并将其应用于存在信息不完整风险的决策评估上。监测诸如桥梁等结构的状况对于确保诸如桥梁等基本公共物品的安全利用至关重要。在这方面，应用人工智能来监测电力工业中必不可少的机电工程结构的状况。

人工智能的一个方面是它如何影响经济理论。一些学者将人工智能应用于股票市场、衍生品和期权等经济金融工具的建模。那么人工智能是如何改变经济理论的呢？例如，诺贝尔经济学奖得主赫伯特·西蒙（Herbert Simon）观察到，在理性地做出决策时，一个人并不具备做出完全理性决策所需的完整信息。此外，一个人没有及时有效地处理这些信息的完美的大脑，人类的大脑是不一致的，因此由人类大脑作出的决定是不一致的，因为它们取决于其他因素，如情绪波动。西蒙把这种情况下的决策称为理性约束。随着人工智能的出现，能够访问原本人类无法触及的隐秘信息，并可以使用这些信息让人工智能进行决策，这些决策将使摩尔定律翻倍。人工智能的出现对有限理性理论意味着什么？这意味着西蒙的有限理性理论中的界限在摩尔定律的作用下是灵活的。

另一个受人工智能影响的经济理论是诺贝尔经济学奖得主尤金·法玛提出的有效率市场假说。这一假说认为，由于市场是有效率的，因此通常很难打败市场规律。问题在于，由于市场上的交易者往往是不完美的，他们所拥有的信息是不完美和不完整的，所以市场是低效率的。如果市场上的交易者不只是人而是人与人工智能的结合，那么有效市场理论会怎样呢？我们在市场上拥有的人工智能赋予计算机交易员的权力越多，市场的效率就越高，因此市场的效率取决于我们在市场上拥有的人工智能交易员的数量。

诺贝尔奖得主丹尼尔·卡尼曼和阿莫斯·特沃斯基提出了前景理论，该理论认为，当人们以已知的结果概率来做决定时，他们会权衡潜在的损失和潜在的收益，从而做出这样的决定。这一理论对市场的影响是广泛的。然而，这取决于一个事实，即决策者只是一个人。如果决策者不只是一个人，而是一个使用人工智能的人，这个理论会发生什么？如果决策者完全是一台人工智能机器呢？前景理论的适用性仅仅取决于有多少人工智能机器被用来做出这样的决定。

决策越来越多地涉及人工智能机制。以上阐述了人工智能机器在决策过程中的应用对三种经济理论的影响。研究发现，人工智能的使用改变了有限理性理论、有效市场假说和前景理论的适用程度。

# Text B

We interpret recent developments in the *field* of artificial intelligence (AI) as improvements in prediction technology. In this paper, we explore the consequences of improved prediction in decision making. To do so, we adapt existing models of decision making under *uncertainty* to account for the process of determining *payoffs*. We label this process of determining the payoffs "judgment." There is a *risky* action, whose payoff depends on the state, and a safe action with the same payoff in every state. Judgment is costly; for each *potential* state, it requires thought on what the payoff might be. Prediction and judgment are complements as long as the expected payoffs in the two states (before judgment is applied) are not too different. We next consider a *tradeoff* between prediction frequency and *accuracy*. We show that as judgment improves, accuracy becomes more important relative to frequency. Finally, we explore the process of gaining experience over time, and show that a seller of predictions cannot *extract* the full value of the predictions from a buyer.

There is widespread discussion regarding the impact of machines on employment, in some sense, the discussion mirrors a long-standing *literature* on the impact of the accumulation of capital *equipment* on employment; specifically, whether capital and labor are substitutes or complements. But the recent discussion is motivated by the *integration* of software with hardware and whether the role of machines goes beyond physical tasks to *mental* ones as well. As mental tasks were seen as always being present and essential, human comparative advantage in these was seen as the main reason why, at least in the long term, capital accumulation would complement employment by enhancing labour productivity in those tasks.

The computer *revolution* has *blurred* the line between physical and mental tasks. For instance, the invention of the spreadsheet in the late 1970s *fundamentally* changed the role of book- keepers. Prior to that invention, there was a time intensive task involving the *recomputation* of outcomes in spreadsheets as data or assumptions changed. That human task was substituted by the spreadsheet software that could produce the calculations more quickly, cheaply, and frequently. However, at the same time, the spreadsheet made

---

**New Words and Expressions**

**field** /fiːld/ n.
　领域；牧场

**uncertainty** /ʌnˈsɜːtnti/ n.
　不确定，不可靠

**payoff** /ˈpeɪɔːf/ n.
　报酬；结果

**risky** /ˈrɪskɪ/ adj.
　危险的；冒险的

**potential** /pəˈtenʃl/ n.
　潜能；可能性

**tradeoff** /ˈtredɔf/ n.
　权衡；折衷

**accuracy** /ˈækjʊrəsɪ/ n.
　精确度，准确性

**extract** /ˈekstrækt/ n.
　提取；取出；摘录；榨取

**literature** /ˈlɪt(ə)rətʃə/ n.
　文学；文献；文艺；著作

**equipment** /ɪˈkwɪpm(ə)nt/ n.
　设备，装备；器材

**integration** /ɪntɪˈɡreɪʃ(ə)n/ n.
　集成；综合

**mental** /ˈment(ə)l/ adj.
　精神的；脑力的

**revolution** /revəˈluːʃ(ə)n/ n.
　革命；旋转；运行；循环

**blur** /blɜː/ vt. vi.
　涂污；使…模糊不清；使暗淡

**fundamentally** /fʌndəˈmentəlɪ/ adv.
　根本地；从根本上；基础地

**recomputation** /ˈriːkəmpjuːˈteɪʃən/ n.
　重新计算；重新估计

the jobs of accountants, analysts, and others far more productive. In the accounting books, capital was substituting for labour but the mental productivity of labour was being changed. Thus, the impact on employment critically depended on whether there were tasks the "computers cannot do."

These assumptions persist in models today. Acemoglu and Restrepo observe that capital substitutes for labour in certain tasks while at the same time technological progress creates new tasks. They make what they call a "natural assumption" that only labour can perform the new tasks as they are more complex than previous ones.1 Benzell, LaGarda, Kotlikoff, and Sachs consider the impact of software more explicitly. Their environment has two types of labour-high-tech (who can, among other things, code) and low-tech (who are empathetic and can handle interpersonal tasks). In this environment, it is the low-tech workers who cannot be replaced by machines while the high-tech ones are employed *initially* to create the code that will *eventually* displace their kind. The results of the model depend, therefore, on a class of worker who cannot be substituted *directly* for capital but also on the *inability* of workers themselves to substitute between classes.

In this paper, our *approach* is to delve into the weeds of what is happening currently in the field of artificial intelligence (AI). The recent wave of developments in artificial intelligence (AI) all involve advances in machine learning. Those advances allow for automated and cheap prediction; that is, providing a forecast (or nowcast) of a variable of interest from available data. In some cases, prediction has enabled full automation of tasks for example, self-driving vehicles where the process of data collection, prediction of behavior and *surroundings*, and actions are all conducted without a human in the loop. In other cases, prediction is a *standalone* tool-such as image recognition or *fraud* detection that may or may not lead to further substitution of human users of such tools by machines. Thusfar, substitution between humans and machines has focused mainly on cost considerations. Are machines cheaper, more reliable, and more scalable (in their software form) than humans? This paper, however, considers the role of prediction in decision-making explicitly and from that examines the complementary skills that may be *matched* with prediction within a take.

For intuition on the difference between prediction and judgment,

**New Words and Expressions**

**initially**/ɪˈnɪʃ(ə)lɪ/ adv.
最初，首先；开头

**eventually**/ɪˈventʃʊəlɪ/ adv.
最后，终于

**directly**/dɪˈrektlɪ; daɪ-/ adv.
直接地；立即；马上；

**inability**/ɪnəˈbɪlɪtɪ/ n.
无能力；无才能

**approach**/əˈprəʊtʃ/ n.
方法；途径；接近

**surrounding**/səˈraʊndɪŋ/ adj.
周围的，附近的

**standalone**/ˈstændəˌləʊn/ n.
独立的

**fraud**/frɔːd/ n.
欺骗；骗子；诡计

**match**/mætʃ/ n. vt. vi.
比赛；匹配

consider the example of credit card fraud. A bank observes a credit card *transaction*. That transaction is either *legitimate* or *fraudulent*. The decision is whether to approve the transaction. If the bank knows *for sure* that the transaction is legitimate, the bank will approve it. If the bank knows for sure that it is fraudulent, the bank will refuse the transaction. Why? Because the bank knows the payoff of approving a legitimate transaction is higher than the payoff of refusing that transaction. Things get more interesting if the bank is uncertain about whether the transaction is legitimate. The uncertainty means that the bank also needs to know the payoff from refusing a legitimate transaction and from approving fraudulent transactions. In our model, judgment is the process of determining these payoffs. It is a costly activity.

As the new developments regarding AI all involve making prediction more readily available, we ask, how does judgment and its *endogenous* application change the value of prediction? Are prediction and judgment *substitutes* or *complements*? How does the value of prediction change *monotonically* with the difficulty of applying judgment? When judgment is a factor, how does this impact on the pricing of AI? Does judgment play a role in the way in which machines learn to predict? And do the answers to these questions change if judgment is an *on-going* activity versus something that can be gained with experience and become *long-lived*?

We *proceed* by first providing *supportive* evidence for our assumption that recent developments in AI *overwhelmingly* impact the costs of prediction. Drawing inspiration from Bolton and Faure-Grimaud, we then build the baseline model with two states of the world and uncertainty about payoffs to actions in each state. We explore the value of judgment in the absence of any prediction technology, and then the value of prediction technology when there is no judgment. We finish the discussion of the baseline model with an exploration of the interaction between prediction and judgment, demonstrating that prediction and judgment are complements as long as the ex ante payoffs to the risky action (before judgment is applied) are expected to be similar. In other words, if the payoffs in the two states are not *anticipated* to be different before judgment is applied, then prediction and judgment are complements. They are substitutes if and only if anticipated differences in payoffs are high

## New Words and Expressions

**transaction**/træn'zækʃ(ə)n/ n.

交易；事务

**legitimate**/lɪ'dʒɪtɪmət/ adj.

合法的；正当的

**fraudulent**/'frɔ:djʊl(ə)nt/ adj.

欺骗性的；不正的

**for sure**

确实；毫无疑问地

**endogenous**/en'dɒdʒɪnəs; ɪn-/ adj.

内生的；内因性的

**substitute**/'sʌbstɪtjuːt/ n.

代用品；代替者

**complement**/'kɒmplɪm(ə)nt/ n.

补足物

**monotonically**/mɒnə'tɒnɪklɪ/ adv.

单调地

**on-going**/ˌɒn'gəʊɪŋ/ adj.

正在进行的；继续的，持续的

**long-lived**/'lɒŋ'lɪvd/ adj.

长命的；历时长久的

**proceed**/prə'siːd/ vi. n.

开始；继续进行

**supportive**/sə'pɔːtɪv/ adj.

支持的；支援的；赞助的

**overwhelmingly**
/ˌəʊvə'hwɛlmɪŋli/ adv.

压倒性地；不可抵抗地

**anticipate**/æn'tɪsɪpeɪt/ vt.

预期，期望

before the decision- maker *invests* in judging the specific payoffs. After these basic results are established, we show that there is no monotonic relationship between improvements in prediction and the value of judgment. We then separate prediction quality into prediction frequency and prediction accuracy. As judgment improves, accuracy becomes more important relative to frequency. Finally, we allow the decision maker to gain experience over time and learn the payoffs given knowledge of the state, without further need to apply judgment. In this *dynamic model*, we show that a seller of predictions (i.e. an AI service provider) cannot extract the full value of the predictions from the buyer.

---

**New Words and Expressions**

**invest**/ɪnˈvest/ vt. vi.

　　投资；覆盖；耗费

**dynamic model**

　　动态模型

---

## 参考译文 B

　　我们将人工智能（AI）领域的最新发展解释为预测技术的进步。本文中我们探讨改进预测对于决策的影响。为此，我们调整了不确定性下的现有决策模型，用来说明确定性收益的流程。我们将这个确定性收益的过程称为"判断"。一种是有风险的行为，其收益取决于所处的状态；另一种是安全的行为，其收益在每种状态下都相同。决策是昂贵的，对于每一种可能的状态，它都需要考虑回报是什么。预测和判断是互补的，只要两种状态（在应用判断之前）的预期收益一致。接下来我们考虑预测频率和准确度之间的关系。我们发现，随着决策能力的提高，准确度相对于频率变得更加重要。最后，我们探索随着时间的推移获得经验的过程，并说明预测的卖方不能从买方那里提取预测的全部价值。

　　关于机器对就业的影响有广泛的讨论，在某种意义上，这种讨论反映了长期以来关于资本设备积累对就业的影响；具体来说，资本和劳动力是替代还是补充。但是，最近软件和硬件的整合，以及机器的作用是否超越了体力劳动和脑力劳动，都推动了新的讨论。由于脑力劳动一直被认为是存在的和必不可少的，在这些任务中，企业通过提高劳动生产率来补充就业，至少在长期内人类是比较有优势的。

　　计算机革命模糊了体力劳动和脑力劳动之间的界限。例如，电子表格在20世纪70年代末的发明从根本上改变了簿记员的角色。在这项发明之前，表格制作是一个时间密集型任务，一旦数据或假设发生变化，则必须重新计算表格中的结果。电子表格软件可以更快、更便宜、更频繁地生成计算，取代了人工劳动。因此，电子表格使会计、分析师和其他人的工作效率大大提高。在会计工作中，机器代替了劳动，但劳动的智力生产力正在发生变化。因此，对就业的影响很大程度上取决于有多少工作是计算机无法完成的。

　　这些假设在今天的模型中仍然存在。Acemoglu 和 Restrepo 观察到，在某些任务中机器替代了劳动力，同时技术进步创造了新的任务。他们做出了他们所谓的"自然假设"，即只有劳动才能完成新任务，因为它们比以前的任务更复杂。Benzell、LaGarda、Kotlikoff 和 Sachs 更明确地考虑了软件的影响。他们的环境有两种类型的劳动力——高科技（除基本事务，还会编码）和低技术（具有同理心，能处理人际关系任务）。在这种环境下，技术含量低的工人无法被机器取代,而高科技工人最初是被雇佣来创建最终将取代他们的代码的。因此, 机器不能直接取代工人, 不同类型的工人之间也不能互相取代。

在本文中, 我们的方法是深入研究人工智能（AI）领域目前正在发生的变革。最近人工智能（AI）的发展都涉及机器学习的进步。这些进步使得自动化和廉价的预测成为可能;也就是说, 根据现有数据提供感兴趣的变量的预测（或短时预测）。在某些情况下, 预测可以实现任务的完全自动化——例如, 自动驾驶汽车的数据收集、行为和环境预测以及行动都是在无人参与的情况下进行的。在其他情况下, 预测是一种独立的工具, 例如图像识别或欺诈检测, 它可能会也可能不会导致机器进一步替代人类用户使用这些工具。因此, 人与机器之间的替代主要集中在成本考虑上。机器（在软件形式上）是否比人类更便宜、更可靠、更灵活? 然而, 本文明确地考虑了预测在决策中的作用, 并从中考察了与预测相匹配的互补技能。

为了直观地理解预测和判断之间的区别, 我们以信用卡欺诈的例子来说明。银行对信用卡交易进行监测。该交易要么合法, 要么具有欺诈性, 并决定是否批准该交易。如果银行确定交易是合法的, 银行就会批准。如果银行确认知道这是欺诈行为, 银行将拒绝交易。为什么?因为银行知道批准合法交易的结果优于拒绝合法交易的结果。如果银行不确定交易是否合法, 事情就会变得更复杂。这种不确定性意味着银行还需要知道拒绝合法交易和批准欺诈性交易的结果。在我们的模型中, 判断是确定这些结果的过程。这是一项非常复杂的活动。

随着人工智能的发展, 预测变得越来越容易, 我们不禁要问, 判断及其应用程序是如何改变预测价值的? 预测和判断是替代还是补充? 预测的价值如何随着判断的难度而变化? 当判断是一个因素时, 这对 AI 的定价有什么影响? 判断在机器学习预测的过程中起作用吗? 如果判断是一项正在进行的活动, 而不是可以通过经验获得并长期存在的东西, 那么这些问题的答案是否会发生变化?

首先为我们的假设提供了支持证据, 即人工智能的最新发展对预测成本产生了压倒性的影响。从 Bolton 和 Faure-Grimaud 那里得到启发, 然后我们构建了基线模型, 其中包含了世界的两种状态以及每个状态中行为回报的不确定性。首先探讨了在没有预测技术的情况下的判断的价值, 然后探讨了在没有预测技术的情况下预测技术的价值。在对基线模型的讨论中, 我们探索了预测和判断之间的相互作用, 证明了预测和判断是互补的, 只要风险行为（在应用判断之前）的事前收益是相似的。换句话说, 如果在应用判断之前, 预期两种状态的回报不存在差异, 那么预测和判断是互补的。当且仅当决策者判断预期的回报差异很大时, 它们才是替代品。在这些基本结果建立了之后, 我们发现预测能力的提高与判断能力的提高之间不存在单调关系。然后将预测质量分为预测频率和预测精度。随着判断能力的提高, 相对于频率, 准确性变得更加重要。最后, 我们允许决策者随着时间的推移获得经验, 并了解给定状态情况下的学习成果, 而不需要进一步进行判断。在这个动态模型中, 我们表明预测的卖家（即 AI 服务提供商）无法从买家那里提取预测的全部价值。

# 附录 *A*

## 常用人工智能词汇中英对照表

**A**

| | |
|---|---|
| Artificial General Intelligence/AGI | 通用人工智能 |
| Artificial Intelligence/AI | 人工智能 |
| association analysis | 关联分析 |
| attention mechanism | 注意力机制 |
| attribute conditional independence assumption | 属性条件独立性假设 |
| attribute space | 属性空间 |
| attribute value | 属性值 |
| autoencoder | 自编码器 |
| automatic speech recognition | 自动语音识别 |
| automatic summarization | 自动摘要 |
| average gradient | 平均梯度 |
| average-pooling | 平均池化 |
| accumulated error backpropagation | 累积误差逆传播 |
| activation Function | 激活函数 |
| Adaptive Resonance Theory/ART | 自适应谐振理论 |
| addictive model | 加性学习 |
| adversarial networks | 对抗网络 |
| affine layer | 仿射层 |
| affinity matrix | 亲和矩阵 |
| agent | 代理 / 智能体 |
| algorithm | 算法 |
| Alpha-Beta pruning | α-β 剪枝 |
| anomaly detection | 异常检测 |

| | |
|---|---|
| approximation | 近似 |
| Area Under ROC Curve/AUC | ROC 曲线下面积 |
| **B** | |
| backpropagation through time | 通过时间的反向传播 |
| BackPropagation/BP | 反向传播 |
| base learner | 基学习器 |
| base learning algorithm | 基学习算法 |
| batch Normalization/BN | 批量归一化 |
| Bayes decision rule | 贝叶斯判定准则 |
| Bayes Model Averaging / BMA | 贝叶斯模型平均 |
| Bayes optimal classifier | 贝叶斯最优分类器 |
| Bayesian decision theory | 贝叶斯决策论 |
| Bayesian network | 贝叶斯网络 |
| between-class scatter matrix | 类间散度矩阵 |
| bias | 偏置/偏差 |
| bias-variance decomposition | 偏差-方差分解 |
| bias-Variance Dilemma | 偏差-方差困境 |
| Bi-directional Long-Short Term Memory/Bi-LSTM | 双向长短期记忆 |
| binary classification | 二分类 |
| binomial test | 二项检验 |
| bi-partition | 二分法 |
| boltzmann machine | 玻尔兹曼机 |
| bootstrap sampling | 自助采样法 / 可重复采样 / 有放回采样 |
| bootstrapping | 自助法 |
| Break-Event Point/BEP | 平衡点 |
| **C** | |
| calibration | 校准 |
| cascade-Correlation | 级联相关 |
| categorical attribute | 离散属性 |
| class-conditional probability | 类条件概率 |
| classification and regression tree/ CART | 分类与回归树 |
| classifier | 分类器 |
| class-imbalance | 类别不平衡 |
| closed -form | 闭式 |
| cluster | 簇/类/集群 |
| cluster analysis | 聚类分析 |

| | |
|---|---|
| clustering | 聚类 |
| clustering ensemble | 聚类集成 |
| co-adapting | 共适应 |
| coding matrix | 编码矩阵 |
| COLT | 国际学习理论会议 |
| committee-based learning | 基于委员会的学习 |
| competitive learning | 竞争型学习 |
| component learner | 组件学习器 |
| comprehensibility | 可解释性 |
| computation Cost | 计算成本 |
| computational Linguistics | 计算语言学 |
| computer vision | 计算机视觉 |
| concept drift | 概念漂移 |
| concept Learning System /CLS | 概念学习系统 |
| conditional entropy | 条件熵 |
| conditional mutual information | 条件互信息 |
| conditional Probability Table/CPT | 条件概率表 |
| conditional random field/CRF | 条件随机场 |
| conditional risk | 条件风险 |
| confidence | 置信度 |
| confusion matrix | 混淆矩阵 |
| connection weight | 连接权 |
| connectionism | 连结主义 |
| consistency | 一致性/相合性 |
| contingency table | 列联表 |
| continuous attribute | 连续属性 |
| convergence | 收敛 |
| conversational agent | 会话智能体 |
| convex quadratic programming | 凸二次规划 |
| convexity | 凸性 |
| convolutional neural network/CNN | 卷积神经网络 |
| co-occurrence | 同现 |
| correlation coefficient | 相关系数 |
| cosine similarity | 余弦相似度 |
| cost curve | 成本曲线 |
| cost Function | 成本函数 |
| cost matrix | 成本矩阵 |
| cost-sensitive | 成本敏感 |
| cross entropy | 交叉熵 |

| cross validation | 交叉验证 |
| crowdsourcing | 众包 |
| curse of dimensionality | 维数灾难 |
| cut point | 截断点 |
| cutting plane algorithm | 割平面法 |

**D**

| data mining | 数据挖掘 |
| data set | 数据集 |
| decision Boundary | 决策边界 |
| decision stump | 决策树桩 |
| decision tree | 决策树/判定树 |
| deduction | 演绎 |
| Deep Belief Network | 深度信念网络 |
| Deep Convolutional Generative Adversarial Network/DCGAN | 深度卷积生成对抗网络 |
| deep learning | 深度学习 |
| Deep Neural Network/DNN | 深度神经网络 |
| Deep Q-Learning | 深度 Q 学习 |
| Deep Q-Network | 深度 Q 网络 |
| density estimation | 密度估计 |
| density-based clustering | 密度聚类 |
| differentiable neural computer | 可微分神经计算机 |
| dimensionality reduction algorithm | 降维算法 |
| directed edge | 有向边 |
| disagreement measure | 不合度量 |
| discriminative model | 判别模型 |
| discriminator | 判别器 |
| distance measure | 距离度量 |
| distance metric learning | 距离度量学习 |
| distribution | 分布 |
| divergence | 散度 |
| diversity measure | 多样性度量 / 差异性度量 |
| domain adaption | 领域自适应 |
| downsampling | 下采样 |
| D-separation（Directed Separation） | 有向分离 |
| dual problem | 对偶问题 |
| dummy node | 哑结点 |
| dynamic Fusion | 动态融合 |
| dynamic programming | 动态规划 |

**E**

| | |
|---|---|
| eigenvalue decomposition | 特征值分解 |
| embedding | 嵌入 |
| emotional analysis | 情绪分析 |
| empirical conditional entropy | 经验条件熵 |
| empirical entropy | 经验熵 |
| empirical error | 经验误差 |
| empirical risk | 经验风险 |
| end-to-End | 端到端 |
| energy-based model | 基于能量的模型 |
| ensemble learning | 集成学习 |
| ensemble pruning | 集成修剪 |
| Error Correcting Output Codes/ECOC | 纠错输出码 |
| error rate | 错误率 |
| error-ambiguity decomposition | 误差-分歧分解 |
| euclidean distance | 欧氏距离 |
| evolutionary computation | 演化计算 |
| Expectation-Maximization | 期望最大化 |
| expected loss | 期望损失 |
| Exploding Gradient Problem | 梯度爆炸问题 |
| exponential loss function | 指数损失函数 |
| Extreme Learning Machine/ELM | 超限学习机 |

**F**

| | |
|---|---|
| factorization | 因子分解 |
| false negative | 假负类 |
| false positive | 假正类 |
| False Positive Rate/FPR | 假正例率 |
| feature engineering | 特征工程 |
| feature selection | 特征选择 |
| feature vector | 特征向量 |
| featured Learning | 特征学习 |
| feedforward Neural Networks/FNN | 前馈神经网络 |
| fine-tuning | 微调 |
| flipping output | 翻转法 |
| fluctuation | 振荡 |
| forward stagewise algorithm | 前向分步算法 |
| Frequentist | 频率主义学派 |
| full-rank matrix | 满秩矩阵 |

| functional neuron | 功能神经元 |
|---|---|

**G**

| gain ratio | 增益率 |
|---|---|
| game theory | 博弈论 |
| Gaussian kernel function | 高斯核函数 |
| Gaussian Mixture Model | 高斯混合模型 |
| general Problem Solving | 通用问题求解 |
| generalization | 泛化 |
| generalization error | 泛化误差 |
| generalization error bound | 泛化误差上界 |
| generalized Lagrange function | 广义拉格朗日函数 |
| generalized linear model | 广义线性模型 |
| generalized Rayleigh quotient | 广义瑞利商 |
| Generative Adversarial Networks/ GAN | 生成对抗网络 |
| Generative Model | 生成模型 |
| generator | 生成器 |
| Genetic Algorithm/GA | 遗传算法 |
| gibbs sampling | 吉布斯采样 |
| Gini index | 基尼指数 |
| global minimum | 全局最小 |
| global Optimization | 全局优化 |
| gradient boosting | 梯度提升 |
| gradient Descent | 梯度下降 |
| graph theory | 图论 |
| ground-truth | 真相/真实 |

**H**

| hard margin | 硬间隔 |
|---|---|
| hard voting | 硬投票 |
| harmonic mean | 调和平均 |
| Hesse matrix | 海塞矩阵 |
| hidden dynamic model | 隐动态模型 |
| hidden layer | 隐藏层 |
| Hidden Markov Model/HMM | 隐马尔可夫模型 |
| hierarchical clustering | 层次聚类 |
| Hilbert space | 希尔伯特空间 |
| hinge loss function | 合页损失函数 |
| hold-out | 留出法 |
| homogeneous | 同质 |

| hybrid computing | 混合计算 |
| --- | --- |
| hyperparameter | 超参数 |
| hypothesis | 假设 |
| hypothesis test | 假设验证 |

**I**

| ICML | 国际机器学习会议 |
| --- | --- |
| improved iterative scaling/IIS | 改进的迭代尺度法 |
| incremental learning | 增量学习 |
| independent and identically distributed/ i.i.d. | 独立同分布 |
| Independent Component Analysis/ ICA | 独立成分分析 |
| indicator function | 指示函数 |
| individual learner | 个体学习器 |
| induction | 归纳 |
| inductive bias | 归纳偏好 |
| inductive learning | 归纳学习 |
| Inductive Logic Programming/ILP | 归纳逻辑程序设计 |
| information entropy | 信息熵 |
| information gain | 信息增益 |
| input layer | 输入层 |
| insensitive loss | 不敏感损失 |
| inter-cluster similarity | 簇间相似度 |
| International Conference for Machine Learning/ICML | 国际机器学习大会 |
| intra-cluster similarity | 簇内相似度 |
| intrinsic value | 固有值 |
| isometric Mapping/Isomap | 等度量映射 |
| isotonic regression | 等分回归 |
| iterative Dichotomiser | 迭代二分器 |

**K**

| kernel method | 核方法 |
| --- | --- |
| kernel trick | 核技巧 |
| Kernelized Linear Discriminant Analysis/KLDA | 核线性判别分析 |
| k-fold cross validation | k 折交叉验证 / k 倍交叉验证 |
| K-Means Clustering | K-均值聚类 |
| K-Nearest Neighbours Algorithm/ KNN | K 近邻算法 |

| knowledge base | 知识库 |
|---|---|
| knowledge Representation | 知识表征 |

**L**

| label space | 标记空间 |
|---|---|
| Lagrange duality | 拉格朗日对偶性 |
| Lagrange multiplier | 拉格朗日乘子 |
| Laplace smoothing | 拉普拉斯平滑 |
| Laplacian correction | 拉普拉斯修正 |
| latent Dirichlet Allocation | 隐狄利克雷分布 |
| latent semantic analysis | 潜在语义分析 |
| latent variable | 隐变量 |
| lazy learning | 懒惰学习 |
| learner | 学习器 |
| learning by analogy | 类比学习 |
| learning rate | 学习率 |
| Learning Vector Quantization/LVQ | 学习向量量化 |
| least squares regression tree | 最小二乘回归树 |
| Leave-One-Out/LOO | 留一法 |
| linear chain conditional random field | 线性链条件随机场 |
| Linear Discriminant Analysis/LDA | 线性判别分析 |
| linear model | 线性模型 |
| linear Regression | 线性回归 |
| link function | 联系函数 |
| local Markov property | 局部马尔可夫性 |
| local minimum | 局部最小 |
| log likelihood | 对数似然 |
| log odds/logit | 对数概率 |
| logistic Regression | Logistic 回归 |
| log-likelihood | 对数似然 |
| log-linear regression | 对数线性回归 |
| Long-Short Term Memory/LSTM | 长短期记忆 |
| loss function | 损失函数 |

**M**

| Machine Translation/MT | 机器翻译 |
|---|---|
| Macron-P | 宏查准率 |
| Macron-R | 宏查全率 |
| Majority Voting | 绝对多数投票法 |
| manifold assumption | 流形假设 |

| | |
|---|---|
| manifold learning | 流形学习 |
| margin theory | 间隔理论 |
| marginal distribution | 边际分布 |
| marginal independence | 边际独立性 |
| marginalization | 边际化 |
| Markov Chain Monte Carlo/MCMC | 马尔可夫链蒙特卡洛方法 |
| Markov Random Field | 马尔可夫随机场 |
| maximal clique | 最大团 |
| maximum Likelihood Estimation/ MLE | 极大似然估计 / 极大似然法 |
| maximum margin | 最大间隔 |
| maximum weighted spanning tree | 最大带权生成树 |
| max-Pooling | 最大池化 |
| mean squared error | 均方误差 |
| meta-learner | 元学习器 |
| metric learning | 度量学习 |
| Micro-P | 微查准率 |
| Micro-R | 微查全率 |
| Minimal Description Length/MDL | 最小描述长度 |
| minimax game | 极小极大博弈 |
| misclassification cost | 误分类成本 |
| Mixture of Experts | 混合专家 |
| momentum | 动量 |
| moral graph | 道德图 / 端正图 |
| multi-class classification | 多分类 |
| multi-document summarization | 多文档摘要 |
| multi-layer feedforward neural networks | 多层前馈神经网络 |
| multilayer Perceptron/MLP | 多层感知器 |
| multimodal learning | 多模态学习 |
| multiple Dimensional Scaling | 多维缩放 |
| multiple linear regression | 多元线性回归 |
| Multi-response Linear Regression/ MLR | 多响应线性回归 |
| mutual information | 互信息 |
| **N** | |
| Naive Bayes | 朴素贝叶斯 |
| Naive Bayes Classifier | 朴素贝叶斯分类器 |
| named entity recognition | 命名实体识别 |

| | |
|---|---|
| Nash equilibrium | 纳什均衡 |
| Natural language generation/NLG | 自然语言生成 |
| natural language processing | 自然语言处理 |
| negative class | 负类 |
| negative correlation | 负相关法 |
| negative Log Likelihood | 负对数似然 |
| Neighbourhood Component Analysis/<br>    NCA | 近邻成分分析 |
| neural Machine Translation | 神经机器翻译 |
| neural Turing Machine | 神经图灵机 |
| Newton method | 牛顿法 |
| NIPS | 国际神经信息处理系统会议 |
| No Free Lunch Theorem/NFL | 没有免费的午餐定理 |
| noise-contrastive estimation | 噪音对比估计 |
| nominal attribute | 列名属性 |
| non-convex optimization | 非凸优化 |
| nonlinear model | 非线性模型 |
| non-metric distance | 非度量距离 |
| non-negative matrix factorization | 非负矩阵分解 |
| non-ordinal attribute | 无序属性 |
| non-Saturating Game | 非饱和博弈 |
| norm | 范数 |
| normalization | 归一化 |
| nuclear norm | 核范数 |
| numerical attribute | 数值属性 |

**O**

| | |
|---|---|
| objective function | 目标函数 |
| oblique decision tree | 斜决策树 |
| Occam's razor | 奥卡姆剃刀 |
| odds | 概率 |
| off-Policy | 离策略 |
| one shot learning | 一次性学习 |
| one-Dependent Estimator/ODE | 独依赖估计 |
| on-Policy | 在策略 |
| ordinal attribute | 有序属性 |
| out-of-bag estimate | 包外估计 |
| output layer | 输出层 |
| output smearing | 输出调制法 |

| | |
|---|---|
| overfitting | 过拟合 / 过配 |
| oversampling | 过采样 |

**P**

| | |
|---|---|
| Paired t-test | 成对 t 检验 |
| pairwise | 成对型 |
| pairwise Markov property | 成对马尔可夫性 |
| parameter | 参数 |
| parameter estimation | 参数估计 |
| parameter tuning | 调参 |
| parse tree | 解析树 |
| particle Swarm Optimization/PSO | 粒子群优化算法 |
| part-of-speech tagging | 词性标注 |
| perceptron | 感知机 |
| performance measure | 性能度量 |
| plug and Play Generative Network | 即插即用生成网络 |
| plurality voting | 相对多数投票法 |
| polarity detection | 极性检测 |
| polynomial kernel function | 多项式核函数 |
| pooling | 池化 |
| positive class | 正类 |
| positive definite matrix | 正定矩阵 |
| post-hoc test | 后续检验 |
| post-pruning | 后剪枝 |
| Potential function | 势函数 |
| precision | 查准率 / 准确率 |
| prepruning | 预剪枝 |
| Principal component analysis/PCA | 主成分分析 |
| principle of multiple explanations | 多释原则 |
| prior | 先验 |
| probability Graphical Model | 概率图模型 |
| proximal Gradient Descent/PGD | 近端梯度下降 |
| pruning | 剪枝 |
| pseudo-label | 伪标记 |

**Q**

| | |
|---|---|
| Quantized Neural Network | 量子化神经网络 |
| quantum computer | 量子计算机 |
| quantum Computing | 量子计算 |
| Quasi Newton method | 拟牛顿法 |

**R**

| | |
|---|---|
| Radial Basis Function / RBF | 径向基函数 |
| Random Forest Algorithm | 随机森林算法 |
| random walk | 随机漫步 |
| recall | 查全率 / 召回率 |
| Receiver Operating Characteristic/ ROC | 受试者工作特征 |
| Rectified Linear Unit/ReLU | 线性修正单元 |
| Recurrent Neural Network | 循环神经网络 |
| recursive neural network | 递归神经网络 |
| reference model | 参考模型 |
| regression | 回归 |
| regularization | 正则化 |
| reinforcement learning/RL | 强化学习 |
| representation learning | 表征学习 |
| representer theorem | 表示定理 |
| reproducing kernel Hilbert space/ RKHS | 再生核希尔伯特空间 |
| Re-sampling | 重采样法 |
| rescaling | 再缩放 |
| residual Mapping | 残差映射 |
| residual Network | 残差网络 |
| Restricted Boltzmann Machine/RBM | 受限玻尔兹曼机 |
| Restricted Isometry Property/RIP | 限定等距性 |
| re-weighting | 重赋权法 |
| robustness | 稳健性/鲁棒性 |
| root node | 根结点 |
| rule Engine | 规则引擎 |
| rule learning | 规则学习 |

**S**

| | |
|---|---|
| saddle point | 鞍点 |
| sample space | 样本空间 |
| sampling | 采样 |
| score function | 评分函数 |
| self-Driving | 自动驾驶 |
| Self-Organizing Map/SOM | 自组织映射 |
| semi-naive Bayes classifiers | 半朴素贝叶斯分类器 |
| semi-supervised Learning | 半监督学习 |
| semi-Supervised Support Vector Machine | 半监督支持向量机 |

| sentiment analysis | 情感分析 |
| separating hyperplane | 分离超平面 |
| sigmoid function | sigmoid 函数 |
| similarity measure | 相似度度量 |
| simulated annealing | 模拟退火 |
| simultaneous localization and mapping | 同步定位与地图构建 |
| Singular Value Decomposition | 奇异值分解 |
| slack variables | 松弛变量 |
| smoothing | 平滑 |
| soft margin | 软间隔 |
| soft margin maximization | 软间隔最大化 |
| soft voting | 软投票 |
| sparse representation | 稀疏表征 |
| sparsity | 稀疏性 |
| specialization | 特化 |
| spectral Clustering | 谱聚类 |
| speech Recognition | 语音识别 |
| splitting variable | 切分变量 |
| squashing function | 挤压函数 |
| stability-plasticity dilemma | 可塑性-稳定性困境 |
| statistical learning | 统计学习 |
| status feature function | 状态特征函 |
| stochastic gradient descent | 随机梯度下降 |
| stratified sampling | 分层采样 |
| structural risk | 结构风险 |
| Structural Risk Minimization/SRM | 结构风险最小化 |
| subspace | 子空间 |
| supervised learning | 监督学习 / 有导师学习 |
| support vector expansion | 支持向量展式 |
| Support Vector Machine/SVM | 支持向量机 |
| surrogat loss | 替代损失 |
| surrogate function | 替代函数 |
| symbolic learning | 符号学习 |
| symbolism | 符号主义 |
| synset | 同义词集 |

**T**

| T-Distribution Stochastic Neighbour Embedding/T-SNE | T-分布随机近邻嵌入 |

| | |
|---|---|
| tensor | 张量 |
| tensor Processing Units/TPU | 张量处理单元 |
| the least square method | 最小二乘法 |
| threshold | 阈值 |
| threshold logic unit | 阈值逻辑单元 |
| threshold-moving | 阈值移动 |
| time Step | 时间步骤 |
| tokenization | 标记化 |
| training error | 训练误差 |
| training instance | 训练示例/训练例 |
| transductive learning | 直推学习 |
| transfer learning | 迁移学习 |
| treebank | 树库 |
| tria-by-error | 试错法 |
| true negative | 真负类 |
| true positive | 真正类 |
| True Positive Rate/TPR | 真正例率 |
| Turing Machine | 图灵机 |
| twice-learning | 二次学习 |

**U**

| | |
|---|---|
| underfitting | 欠拟合/欠配 |
| undersampling | 欠采样 |
| understandability | 可理解性 |
| unequal cost | 非均等代价 |
| unit-step function | 单位阶跃函数 |
| univariate decision tree | 单变量决策树 |
| unsupervised learning | 无监督学习/无导师学习 |
| unsupervised layer-wise training | 无监督逐层训练 |
| upsampling | 上采样 |

**V**

| | |
|---|---|
| vanishing Gradient Problem | 梯度消失问题 |
| variational inference | 变分推断 |
| VC Theory | VC 维理论 |
| version space | 版本空间 |
| viterbi algorithm | 维特比算法 |
| Von Neumann architecture | 冯·诺依曼架构 |

**W**

| | |
|---|---|
| Wasserstein GAN/WGAN | Wasserstein 生成对抗网络 |
| weak learner | 弱学习器 |

| | |
|---|---|
| weight | 权重 |
| weight sharing | 权共享 |
| weighted voting | 加权投票法 |
| within-class scatter matrix | 类内散度矩阵 |
| word embedding | 词嵌入 |
| word sense disambiguation | 词义消歧 |

**Z**

| | |
|---|---|
| zero-data learning | 零数据学习 |
| zero-shot learning | 零次学习 |